AF361964

Energy-Water Nexus

Energy-Water Nexus

Editors

Antonio Colmenar Santos
David Borge Diez
Enrique Rosales Asensio

MDPI • Basel • Beijing • Wuhan • Barcelona • Belgrade • Manchester • Tokyo • Cluj • Tianjin

Editors

Antonio Colmenar Santos
Department of Electrical,
Electronic, Control, Telematics
and Chemical Engineering
Applied to Engineering,
Higher Technical School of
Industrial Engineers,
National University of
Distance Education
Spain

David Borge Diez
Energy Resources' Smart
Management (ERESMA)
Research Group, Department
Area of Electrical Engineering,
School of Mines Engineering,
University of Léon
Spain

Enrique Rosales Asensio
Department of Physics,
University of La Laguna
Spain

Editorial Office
MDPI
St. Alban-Anlage 66
4052 Basel, Switzerland

This is a reprint of articles from the Special Issue published online in the open access journal *Energies* (ISSN 1996-1073) (available at: https://www.mdpi.com/journal/energies/special_issues/energy_water_nexus_2020).

For citation purposes, cite each article independently as indicated on the article page online and as indicated below:

LastName, A.A.; LastName, B.B.; LastName, C.C. Article Title. *Journal Name* **Year**, *Volume Number*, Page Range.

ISBN 978-3-0365-0084-3 (Hbk)
ISBN 978-3-0365-0085-0 (PDF)

Cover image courtesy of NASA, public domain CC0 image.

Contents

About the Editors

Antonio Colmenar Santos has been a senior lecturer in the field of Electrical Engineering at the Department of Electrical, Electronic and Control Engineering at the National Distance Education University (UNED) since June 2014. Dr. Colmenar-Santos was an adjunct lecturer at both the Department of Electronic Technology at the University of Alcalá and at the Department of Electric, Electronic and Control Engineering at UNED. He has also worked as a consultant for the INTECNA project (Nicaragua). He has been part of the Spanish section of the International Solar Energy Society (ISES) and of the Association for the Advancement of Computing in Education (AACE), working in a number of projects related to renewable energies and multimedia systems applied to teaching. He was the coordinator of both the virtualisation and telematic Services at ETSII-UNED, and deputy head teacher and the head of the Department of Electrical, Electronics and Control Engineering at UNED. He is the author of more than 60 papers published in respected journals (http://goo.gl/YqvYLk) and has participated in more than 100 national and international conferences.

David Borge Diez has a Ph.D. in Industrial Engineering and an M.Sc. in Industrial Engineering, both from the School of Industrial Engineering at the National Distance Education University (UNED). He is currently a lecturer and researcher at the Department of Electrical, Systems and Control Engineering at the University of León, Spain. He has been involved in many national and international research projects investigating energy efficiency and renewable energies. He has also worked in Spanish and international engineering companies in the field of energy efficiency and renewable energy for over eight years. He has authored more than 40 publications in international peer-reviewed research journals and participated in numerous international conferences.

Enrique Rosales Asensio is an industrial engineer with postgraduate degrees in electrical engineering, business administration, and quality, health, safety and environment management systems. He has been a lecturer at the Department of Electrical, Systems and Control Engineering at the University of León, and a senior researcher at the University of La Laguna, where he has been involved in water desalination project in which the resulting surplus electricity and water would be sold. He has also worked as a plant engineer for a company that focuses on the design, development and manufacture of waste-heat recovery technology for large reciprocating engines, and as a project manager in a world-leading research centre. Currently, he is an associate professor at the Department of Electrical Engineering at the University of Las Palmas de Gran Canaria.

Preface to "Energy-Water Nexus"

Water is necessary to produce energy, and energy is required to pump, treat, and transport water. The energy–water nexus examines the interactions between these two inextricably linked elements. This Special Issue aims to explore a single "system of systems" for the integration of energy systems. This approach considers the relationships between electricity, thermal, and fuel systems; and data and information networks in order to ensure optimal integration and interoperability across the entire spectrum of the energy system. This framework for the integration of energy systems can be adapted to evaluate the interactions between energy and water. This Special Issue focuses on the analysis of water interactions with and dependencies on the dynamics of the electricity sector and the transport sector.

Antonio Colmenar Santos, David Borge Diez, Enrique Rosales Asensio
Editors

Article

Carbon and Water Footprint of Energy Saving Options for the Air Conditioning of Electric Cabins at Industrial Sites

Maurizio Santin [1], Damiana Chinese [1],*, Onorio Saro [1], Alessandra De Angelis [1] and Alberto Zugliano [2]

[1]   Dipartimento Politecnico di Ingegneria e Architettura (DPIA), University of Udine, Via delle Scienze 206, 33100 Udine (UD), Italy; santin.maurizio@spes.uniud.it (M.S.); onorio.saro@uniud.it (O.S.); alessandra.deangelis@uniud.it (A.D.A.)

[2]   Danieli & C. Officine Meccaniche S.p.A., Via Nazionale, 41, 33042 Buttrio (UD), Italy; a.zugliano@danieli.it

*   Correspondence: damiana.chinese@uniud.it; Tel.: +39-0432-558024

Received: 26 August 2019; Accepted: 19 September 2019; Published: 23 September 2019

Abstract: Modern electric and electronic equipment in energy-intensive industries, including electric steelmaking plants, are often housed in outdoor cabins. In a similar manner as data centres, such installations must be air conditioned to remove excess heat and to avoid damage to electric components. Cooling systems generally display a water–energy nexus behaviour, mainly depending on associated heat dissipation systems. Hence, it is desirable to identify configurations achieving both water and energy savings for such installations. This paper compares two alternative energy-saving configurations for air conditioning electric cabins at steelmaking sites—that is, an absorption cooling based system exploiting industrial waste heat, and an airside free-cooling-based system—against the traditional configuration. All systems were combined with either dry coolers or cooling towers for heat dissipation. We calculated water and carbon footprint indicators, primary energy demand and economic indicators by building a TRNSYS simulation model of the systems and applying it to 16 worldwide ASHRAE climate zones. In nearly all conditions, waste-heat recovery-based solutions were found to outperform both the baseline and the proposed free-cooling solution regarding energy demand and carbon footprint. When cooling towers were used, free cooling was a better option in terms water footprint in cold climates.

Keywords: waste heat recovery; absorption cooling; water–energy nexus; steelworks; TRNSYS

---

1. Introduction

The iron and steelmaking industry is an energy-intensive sector that accounts for about 18% of the world's total industry final energy consumption [1]. Steelmaking processes are also carbon intensive, and the sector accounts for 5% of global CO_2 emissions [2].

Consequently, the steelmaking industry is currently subjected to emission trading schemes (ETSs) in several countries [3,4]. Overall, emission certificate costs have been low in recent years, hardly providing steel plant operators with an economic rationale to reduce their energy demand and emissions. However, progressively more stringent environmental standards and energy policy scenarios increase the likelihood of a rise in primary energy and CO_2-emission certificate costs [5]. To avoid a consequent increase in the market price of steel products, it is crucial for steelmaking industries to identify cost-effective solutions for carbon emission reduction.

Worldwide steelmaking industries are also aware of the water–energy nexus [6] implications of their attempts to improve efficiency: a position paper on water saving by the World Steel Association [7]

points out that "the additional processes (required to save water) are nearly always in conflict with objectives to reduce energy consumption or CO_2 emissions".

Similarly, research in the steel sector also reports some unexpected cases of water consumption increase as an observed outcome of energy-saving measures in real settings [8] or as a potential consequence of suboptimal carbon reduction practices under simulated incentive frameworks [9]. This may even happen in the case of waste-heat recovery [9,10], which is generally considered a synergistic option to decrease water and energy demand, as far as it reduces the need to discard water into the environment via cooling towers and cooling fans [11]. Therefore, to highlight synergies and avoid pitfalls, it is important that energy-saving projects in steelmaking and, more generally, in energy- and water-intensive industries, are evaluated with a nexus view [6], considering their impact on primary energy consumption and carbon emissions, as well as on water consumption.

Overviews of heat recovery options in the steelmaking industry have been presented by Moya and Pardo [12], He and Wang [1], as well as Johansson and Söderström [13]. Several waste heat utilization practices have been proposed, including iron-ore or scrap pre-heating, in basic oxygen furnace (BOF) steelmaking cycles, or electric-arc furnaces (EAFs), respectively, as well as power generation with Rankine cycles [13] which to date mainly occurs in BOF plants [14].

However, all these waste-heat utilization routes allow the exploitation of only a fraction of the sizeable waste heat flows available at steelmaking sites [15]. To improve energy efficiency and decarbonize industries, other forms of the utilization of waste heat are sought, particularly as direct or upgraded heat utilization [13].

One option for the internal utilization of medium-low-temperature waste heat flows in steelmaking, and more generally for energy-intensive industries, is to identify some process cooling demand that is currently met with vapour compression cooling systems, and substitute these with waste-heat-based absorption cooling systems. In fact, absorption cooling is a mature technology [16] that makes use of low global warming potential and non-ozone-layer-depleting natural materials as working fluid pairs. In particular, H_2O–LiBr and NH_3–H_2O are the best performing and most common working fluid pairs [17]. H_2O–LiBr systems are safer and less complex than NH_3–H_2O systems; the latter are therefore almost exclusively used for applications requiring refrigeration temperatures below 0 °C. Overall, absorption cooling running on solar heat or on waste heat sources can be regarded as zero-carbon-emission cooling systems [18].

For absorption cooling based air conditioning systems, the literature has focused primarily on solar cooling [19]. Most solar cooling applications make use of single effect cycles, which can be regarded as the state-of-the-art commercially mature technology for low-temperature applications running on hot water below 100 °C [16]. With the increasing spread of parabolic concentrators, double, triple, and variable effect cycles have been increasingly investigated [19,20], as they are more adept at exploiting medium-temperature heat sources (up to about 260 °C; see [18]) by enabling systems to reach the coefficient of performance (COP; or energy efficiency ratio (EER) on the order of 1.25 (double effect [18]) or 2 (triple effect [18]), depending on the heat source temperature, whereas single effect cycles have EERs on the order of 0.7 (hot water temperature on the order of 90 °C [21]). Readers are asked to bear in mind that in this paper we will refer to this parameter as EER, in accordance with the terminology introduced by standard EN 14511, which defines the EER as the ratio of the total cooling capacity of refrigerators to their effective power input, both expressed in Watt.

Practical industrial waste heat (IWH)-based cooling applications have received relatively less attention than solar cooling in the literature: the technology was proposed in some review papers [22,23], and mathematical models for the optimization of district cooling applications based on industrial waste heat recovery have recently been proposed for illustrative case studies from the chemical industry [24,25]. Some techno-economic feasibility assessments of absorption cooling as a recovery option for industrial low-grade waste heat have been performed by Brückner et al. [26] for general European IWH potentials, and by Cola et al. [27] for a drying process in the textile industry. In both cases, the assessment was performed either on a purely economic [26] basis or on an economic and thermodynamic basis [27]. However, the environmental implications of different choices, particularly with a water–energy-nexus-aware view, have hardly been considered.

An application of H_2O–LiBr single effect absorption cooling to the air conditioning of electric transformer, generator, and switch cabins for the steelmaking industry was recently proposed in [9]. In fact, electric cabins must be air conditioned to remove excess heat and avoid damage to electric components, in order to avoid abnormal functioning or breakdowns of electric equipment due to Joule heating. This is especially true for EAF steelmaking sites, where transformers are required to provide electricity to all the electric equipment (e.g., electric motors, control rooms, robots, and the EAF electrodes). However, this application could be of interest for any industrial site housing large transformers in electric cabins.

The authors of [9] demonstrated that at average climate conditions for the EU-15 area, absorption cooling is economically preferable to Organic Rankine Cycle (ORC)-based power generation for exploiting intermittent low-grade waste heat flows available at EAF steelmaking sites. Moreover, they performed an assessment of the carbon emission and water consumption performance of those systems at average EU conditions.

However, they admit that a limitation of their study is that climate differences among different countries have not been considered, assuming a constant cooling demand for the whole year and for the entire area of analysis.

This may be acceptable when considering cabins located within industrial sheds and when performing comparisons for geographically limited areas. However, modern electric and electronic equipment in energy-intensive industries, including electric steelmaking plants, is often housed in outdoor cabins. The water–energy impact of such systems is likely to be affected by climate, particularly depending on the residual waste heat dissipation systems installed, such as forced air coolers (a.k.a. dry coolers, DCs in the following) or cooling towers (CTs). The energy and money savings generated by heat-recovery-based cooling systems might even be negligible in some climates, and other options might be more efficient for cabin air conditioning.

The present study aims to overcome the mentioned limitations, and to investigate the economic and water–energy nexus implications of exploiting low-grade process waste heat in outdoor electric cabins worldwide, based on typical situations at EAF sites.

To the best of the authors' knowledge, this problem has not yet been addressed on this scale. However, some input for research design and methodology selection could be obtained from research on data centres [28–32], which also need intensive and continuous cooling to preserve electric and electronic components. Indeed, for data centres, absorption cooling has been proposed as a means to recover waste heat from the internal electric equipment (e.g., a subset of servers) to meet a part of internal cooling loads [28,29]. However, to the best of the authors' knowledge, the opportunity of exploiting an external waste heat source to feed absorption cooling systems for data centre air conditioning has not been investigated. On the other hand, direct air free cooling technology, which uses the cold outside air to remove the heat generated inside these facilities, has been extensively investigated for data centres [30–32], and could be an interesting low-cost option for electric cabins as well.

From a water–energy nexus perspective, this paper aims to determine whether and where process waste heat recovery for absorption cooling may be a better option than airside free cooling for maintaining acceptable temperatures within electric cabins. Thereby, this research is expected to widen current knowledge of the environmental performance of absorption cooling systems as a recovery option for low-grade industrial waste heat, particularly from a water–energy nexus perspective.

2. Methodology

To achieve the objectives mentioned above, a reference electric cabin is defined in Section 2.1, and the air conditioning configurations described in Section 2.2 were examined—that is, traditional vapour compression cooling (baseline strategy, mechanical vapour compression (MVC)), vapour compression cooling combined with airside free cooling (FC), and waste-heat-recovery-based absorption cooling (ABS). Each air conditioning option was evaluated in combination with either DC or CT in order to identify the best-performing configurations.

The cooling systems for the reference cabin were modelled with the transient energy simulation software TRNSYS [33] in locations representing worldwide climate zones as defined by the ASHRAE [34] using climate data available with TRNSYS, as specified in Section 2.3. To evaluate the water–energy impact of these systems, the primary energy demand as well as the carbon and water footprints were calculated for each configuration by evaluating the direct electricity and water consumption based on simulations, as well as indirect contributions such as carbon emissions, primary energy and blue water consumption associated with electricity generation in each location, based on the approach and assumption described in Section 2.4. The economic efficiency was also assessed using the data reported in Section 2.5, particularly by establishing if and where absorption cooling is able to compete with the airside free cooling configuration.

2.1. Air Conditioning System and Building Specifications

The cabin cooling system consists of an air-cooling unit located inside the room, where the thermostat is set to keep the inside temperature under 40 °C—a safety operation threshold provided by electric equipment manufacturers.

Compared with data centres [30], the regulation requirements for electric cabins at steelmaking sites are substantially less restrictive, as they house robust equipment designed for harsh working environments. Thus, in this study, it was assumed that the temperature control system operates with a set point temperature of 35 ± 2.5 °C. In this analysis, a 1000-kW cooling load from internal equipment was assumed as typical for a reference electric cabin having a building surface area of 3700 m^2 and a volume of 17,000 m^3. Outside electric cabins were investigated in the present work in order to determine the extent to which local climate affects the cabin cooling load and the performance of different cooling systems. The thermal transmittance of the cabin was evaluated based on data provided by cabin manufacturers at 0.4 W/m^2·K.

2.2. Cooling Systems Configurations

Three alternative cabin air conditioning configurations are modelled and compared in this study: the baseline mechanical vapour compression chiller (MVC) described in Section 2.2.1; an energy-saving mechanical vapour compression configuration based on airside free cooling with outside air (FC), presented in Section 2.2.2; and a waste heat recovery absorption-cooling-based configuration (ABS), as specified in Section 2.2.3. In particular, as in [5] and [9], it is proposed to recover waste heat from the hot gas line cooling system of conventional electric arc furnaces based on the plant layout and temperature profiles reported in [9]. In fact, in conventional EAFs, off-gases leaving the furnace and the following dropout box are cooled down to at least 600 °C, as required for the operation of subsequent plant components, by flowing through a modular gas-tight water-cooled duct [5,35], known in the industry as a WCD. In conventional configurations, the water used as refrigerant in the WCD needs to be cooled down in heat rejection units (i.e., either DC or CT). Total removed heat loads vary over time due to process intermittence, and depending on steelworks capacity, reaching values ranging between 10 and 20 MW for a 130 t nominal tap weight furnace [36]. For the heat recovery system of concern, we considered the opportunity to derive a water flow from a module of the cooling water circuit corresponding to an average heat flow of about 3100 kW. To obtain a simple and homogenous assessment of the impact of heat rejection units depending on climate, it was assumed that the same technology (i.e., either DC or CT) was used both for heat rejection at the WCD and as condenser for cabin refrigeration cycles.

2.2.1. Water-Cooled MVC Chiller

Mechanical vapour compression chillers are the most common refrigerators for air-conditioning purposes. In this study a water-cooled magnetic centrifugal chiller was selected as baseline refrigeration system for electric cabin air conditioning. The nominal capacity installed was 1300 kW and the performance was taken from a York catalogue for chillers [37]. The EER was 6.4, evaluated at an entering/leaving chilled water temperature of 12/7 °C and entering/leaving condenser water temperature of 30/35 °C.

Figure 1 shows the scheme of this configuration, which depicts both the cabin air conditioning system and the module of the WCD cooling circuit selected for heat recovery in configuration 2.2.3.

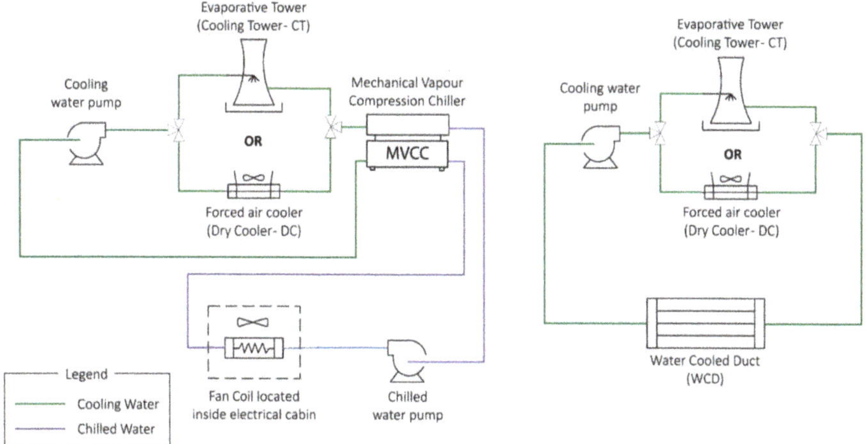

Figure 1. Mechanical vapour compression chiller schematic diagram.

2.2.2. Free Cooling and MVC chiller

The FC configuration analysed in this paper, represented in Figure 2, consists of an MVC air conditioning configuration coupled with an external air ventilation system which draws air from outside and, after filtering, directly introduces it into the cabin, thereby reducing the cooling load for the conventional MVC chiller. In order to reduce the computational load without losing the significance in comparison, a fixed value of external air temperature was chosen to control the operation of the free cooling system. A value of 18 °C was assumed as the switch-off temperature, allowing the capacity of the free cooling ventilation system to be comparable with the internal fans' capacity. Thus, when the external air temperature is higher than 18 °C, the standard MVC chiller operates to cool the internal cabin air. Otherwise, the system operates in FC mode. Also in this case, no heat recovery from the WCD occurs and its full load is dissipated at heat rejection units.

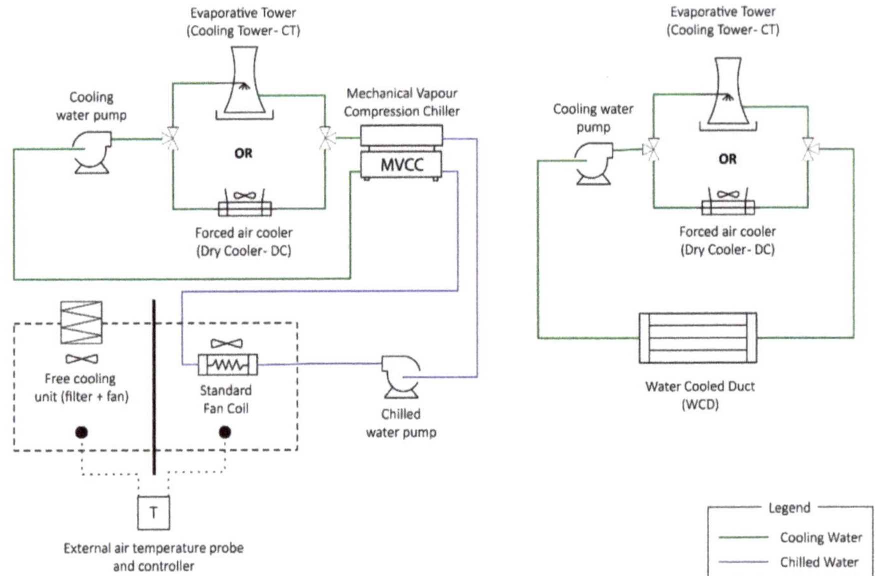

Figure 2. Mechanical vapour compression chiller with free cooling system schematic diagram.

2.2.3. Air-Cooler and Water-Cooled ABS Chiller

The waste-heat-recovery-based cooling system represented in Figure 3 relies on a hot-water-fed single effect absorption chiller. As underlined in [5], in conventional WCDs at EAFs, due to no further utilization purposes of the emitted thermal energy, the cooling water outlet temperature is usually in the range of 50 °C [38]. If thermal energy recovery is considered, the design temperature of the cooling system has to be increased.

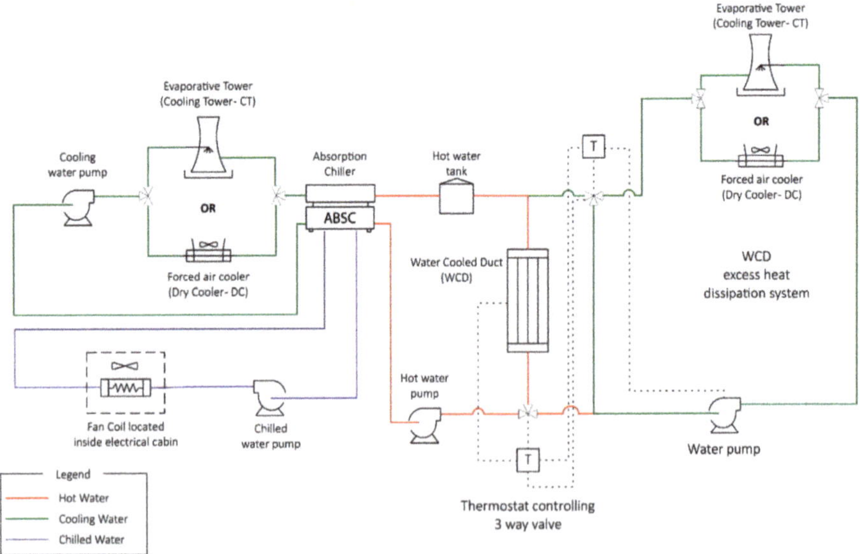

Figure 3. Absorption chiller schematic diagram.

While it is also feasible to increase it to 200 °C, as demonstrated in [5], for this absorption cooling application the choice was made to increase it only to the average value of 90 °C. In this way, the system was designed to operate with hot water, in order to avoid introducing additional complexities from steam operation, such as additional maintenance and safety requirements related to higher temperature, pressure and phase change, which would be an additional burden in EAF plants without or with minimal steam networks. With hot water, single effect absorption chillers are used, whose reference EER is in the order of 0.7, in accordance with manufacturers' catalogues [39,40] and the literature [21]. A commercial absorption cooling system with a nominal capacity of 1319 kW was assumed to be installed, based on the LG Absorption Chiller catalogue [40]. At EAF steelmaking sites where steam networks exist, an integrated development of heat-recovery-based steam generation as in [5] and of absorption-based cooling could be considered in order to exploit more efficient double effect absorption cycles [18,26]. However, this is beyond the scope of the present paper. Given the intermittence of the EAF melting process, based on the aforementioned tap-to-tap cycle, variations in flue gas temperatures correspond to oscillations in cooling water temperature at the heat recovery outlet. Thus, as in [5] and [9] a water storage tank is used as a hot water reservoir to compensate for power-off phases by limiting the temperature variability, which for single effect absorption cooling purposes is deemed acceptable in the range of 85 to 95 °C. The hot storage size was also designed to meet safety design criteria for cabin air conditioning systems, which imply that the cooling load to be removed from electric cabins was assumed to be constantly present during steelworks operations and to persist, during maintenance stops, for a period of three hours after the steelworks stop.

2.3. TRNSYS Simulation Model Development

TRNSYS [33] was used in this work to perform a dynamic simulation of the behaviour of the elements used in the various configurations analysed. TRNSYS libraries consist of components such as heating, ventilation and air conditioning (HVAC), electronics, controls, hydronics, etc. The elements are called types, and can be linked to others to simulate entire systems.

Dynamic system simulation is possible by including performance data and simulation parameters for individual elements. The configurations defined in Section 2 were modelled in TRNSYS based on the schematic diagrams shown in Figures 1–3, obtaining TRNSYS input files (usually referred to as decks). As an example, the TRNSYS deck for the ABS configuration is represented in Figure 4.

The mass flowrates of chilled water and condenser water required by chillers were taken from manufacturers' catalogues. MVC and ABS systems were simulated using technical data (reference chilled, cooling, and hot water flow rates) from the manufacturers' catalogues [37] and [40], respectively, and the TRNSYS inbuilt performance data file, which allows EER simulation as a function of cooling, chilled, and hot water temperatures. The hot water tank was a stratified, five-layer adiabatic liquid storage tank simulated using TRNSYS *type60*. Cooling water from CTs or DCs served as input for the chiller condenser while the water leaving the chiller condenser was used as the input in CTs or DCs depending on the configuration studied. For the simulation of heat rejection units, technical data required as TRNSYS input were taken from LU-VE catalogue [41] and YWCT catalogue [42] for DCs and CTs, respectively.

A weather data file derived from the EnergyPlus weather database [43] was provided as input to CT, DC, and the cabin building to capture the effect of external environment conditions. Climate zones were selected according to ASHRAE [34]. Table 1 shows the cities selected here to represent each climate zone and their climate characteristics. The three cooling configurations described in the previous section were simulated in 16 out of 17 of the selected cities. Climate zone number 8 was not considered for simulations since cooling towers are inoperable in this zone [32] due to the extremely low temperatures (see Table 1). Therefore, a total of 96 simulations were performed.

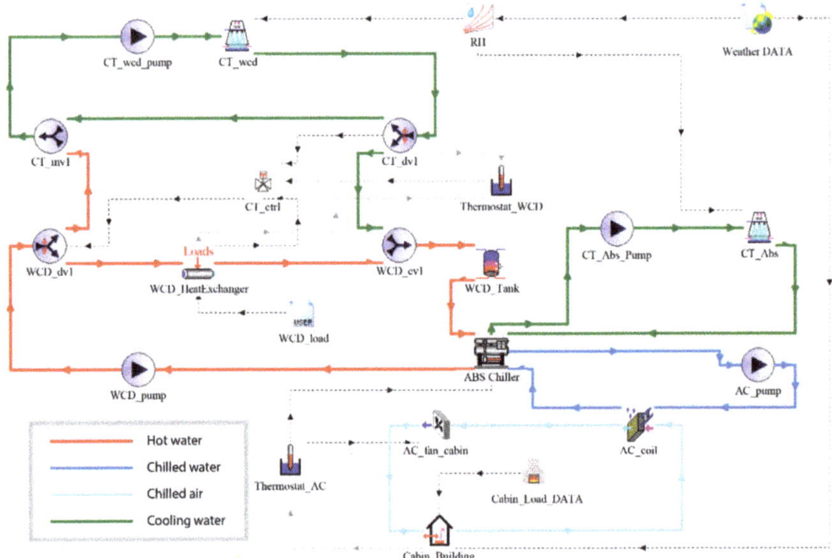

Figure 4. Absorption cooling (ABS) configuration modelled in TRNSYS. Dashed lines are used as control indicators.

Table 1. Climate zones defined by ASHRAE and relative representative city. Zone number 8 (in italics) was not considered in this work.

Climate Zone	City	DRY BULB t. (°C)			WET BULB t. (°C)			RH (%)		
		Min	Max	Mean	Min	Max	Mean	Min	Max	Mean
1A	Singapore	21.1	33.8	27.5	16.9	28.2	25.1	44	100	84
1B	New Delhi	5.2	44.3	24.7	4	29.5	19	9	99	62
2A	Taipei	6	38	22.8	5.1	29	20.3	35	100	81
2B	Cairo	7	42.9	21.7	6	27	15.9	10	100	59
3A	Algiers	−0.8	38.5	17.7	-1	27.1	14.6	13	100	75
3B	Tunis	1.3	39.9	18.8	1.2	26.8	15.2	14	100	72
3C	Adelaide	2	39.2	16.2	1.2	25.2	11.7	6	100	63
4A	Lyon	−8.5	33.6	11.9	−9.2	26.2	9.4	16	100	76
4B	Seoul	−11.8	32.7	11.9	−13.3	29.6	9.2	9	100	69
4C	Astoria	−3.3	28.3	10.3	−4.7	21.4	8.6	29	100	81
5A	Hamburg	−8.5	32	9	−9.2	22.8	7.1	26	100	80
5B	Dunhuang	−19.6	39.1	9.8	−20	24.3	3.6	4	98	42
5C	Birmingham	−7.4	30.4	9.7	−7.8	20.3	7.7	19	100	78
6A	Moscow	−25.2	30.6	5.5	−25.2	21.7	3.7	28	100	77
6B	Helena	−29.4	36.1	6.8	−29.7	19.1	2.5	11	100	57
7	Ostersund	−25.7	26.5	3.2	−26.1	18.5	1.3	23	100	75
8	*Yakutsk*	*−48.3*	*32.1*	*−9.1*	*−48.3*	*20*	*−11.1*	*14*	*100*	*68*

2.4. Calculation of Water–Energy–Greenhouse Gas (GHG) Nexus Indicators

In accordance with [9,44], the total blue water footprint, carbon footprint and primary energy demand were selected as water–energy–carbon nexus indicators in this analysis. They meet most requirements reported by [45,46] for sustainability indicators; in particular, they are easy to interpret, able to show trends over time and sensitive to changes in the systems analysed here (i.e., different configurations of cabin refrigeration systems).

2.4.1. Water Footprint

The total water footprint W_f was calculated as the sum of the water consumption within systems (direct water use, W_d) and the water footprint of energy consumed (indirect water use, W_{ind}) according to Equation (1):

$$W_f = W_d + W_{ind} = k \cdot W_{ev} + C_{W,el} E_{el}. \tag{1}$$

In the present evaluation, we did not account for water pollution impacts (so-called grey water), but only for blue water footprint, which measures the consumptive use of surface and ground water.

Direct water consumption only occurs in CT configurations due to evaporation loss, drift and makeup-water requirements. Evaporated quantities were calculated with TRNSYS [33] using *type51b*. Additional water losses due to bleed off and drift were quantified as in [9] using a multiplicative coefficient k on the evaporated water W_{ev}, taking $k = 2$ as a reasonable estimate [47]. The footprint calculation approach and the data sources reported in [9] were used to derive the indirect water consumption rate $C_{W,el}$ for each reference city based on the national electricity production mix reported in Table 2, elaborated from the WorldBank database [48].

The total electricity demand E_{el} was determined as the sum of the energy required for each component simulated in TRNSYS. The chiller performance was considered in the energy consumption calculation by using the corresponding TRNSYS types. In-built TRNSYS performance data files were used to evaluate the EER and consequently the energy consumption, which is related to the cooling water temperature returning from the heat rejection device (DC or CT) as well as the temperature of chilled outlet water. For the absorption chiller, the inlet hot water temperature was also introduced as parameter to determine the EER. As a result, the yearly average EER values obtained from simulations in the climate regions of concern ranged between 0.52 and 0.55 for absorption cooling systems, and between 5.24 and 9.61 for compression cooling systems.

2.4.2. Carbon Footprint and Primary Energy Demand Calculation

Carbon footprint has been defined as "the quantity of GHGs expressed in terms of CO_2 equivalent mass emitted into the atmosphere by an individual, organization, process, product or event from within a specified boundary" [49].

As in the case of water footprint, differences in the carbon footprint of the configurations examined are exclusively bound to electricity consumption, since none of the air conditioning alternatives examined implies any direct fuel consumption. Carbon footprint was thus calculated according to Equation (2).

$$CO2_f = CO2_{ind} = C_{CO2,el} E_{el}. \tag{2}$$

On the other hand, based on the data sources used in this study (see [9]), carbon footprint coefficients for electricity consumption $C_{CO2,el}$ were estimated with a life cycle approach (i.e., all $CO2_{eq}$ emissions consumption from extraction to plant construction were considered).

In a similar manner to [50], in this study it was assumed that the changes in direct carbon equivalent emissions from refrigerant leaks induced by switching from vapour compression units to absorption cooling systems were negligible compared to the emissions of greenhouse gases embodied in purchased electricity.

The primary energy consumption associated with purchased electricity was calculated according to Equation (3):

$$PED = C_{PED,el} E_{el}. \tag{3}$$

Site-to-source energy conversion factors $C_{PED,el}$ reported in Table 2 were obtained with the methodology and data sources discussed in [9,51] based on national energy mix data reported in Table 2.

Table 2. National energy mix for the analysed countries (climates) in this paper [44]. TOE: ton (of) oil equivalent.

City	Nation	Biomass and Waste	Solid Fuels	Natural Gas	Geothermal Energy	Hydropower	Nuclear Energy	Crude Oil	Solar Energy	Wind Energy	$C_{CO2,el}$ (tCO$_2$/GWh)	$C_{W,el}$ (m^{3}H$_2$O/GWh)	$C_{PED,el}$ (TOE/GWh)
Singapore	Singapore	1.39%	0.00%	79.77%	0.00%	0.00%	0.00%	18.82%	0.02%	0.00%	524	1105	212
New Delhi	India	0.51%	68.21%	10.35%	0.96%	12.69%	3.02%	1.17%	0.21%	2.88%	745	6284	228
Taipei	Taiwan	1.45%	32.65%	10.88%	0.00%	2.40%	16.55%	35.32%	0.01%	0.73%	646	2510	254
Cairo	Egypt	0.00%	0.00%	74.59%	0.00%	8.70%	0.00%	15.73%	0.15%	0.83%	478	4189	196
Algiers	Algeria	0.00%	0.00%	93.39%	0.00%	1.13%	0.00%	5.48%	0.00%	0.00%	490	1463	202
Tunis	Tunisia	0.00%	0.00%	98.19%	0.00%	0.62%	0.00%	0.00%	0.00%	1.18%	473	1254	197
Adelaide	Australia	0.98%	69.33%	19.28%	0.00%	5.82%	0.00%	1.41%	0.63%	2.56%	799	3754	235
Lyon	France	0.98%	3.99%	3.60%	0.00%	10.89%	76.33%	0.57%	0.84%	2.79%	82	5909	245
Seoul	South Korea	0.24%	42.36%	23.03%	0.00%	0,79%	29.03%	4.14%	0.22%	0.18%	573	2094	254
Astoria	USA	1.77%	38.23%	29.77%	0.09%	6.84%	19.04%	0.68%	0.11%	3.48%	538	4057	227
Hamburg	Germany	7.68%	46.37%	11.33%	0.00%	3.61%	16.20%	1.53%	4.54%	8.73%	541	2973	221
Dunhuang	China	0.95%	74.94%	1.69%	0.00%	18.14%	1.96%	0.16%	0.13%	2.03%	764	8226	223
Birmingham	UK	4.23%	40.09%	27.84%	0.14%	1.55%	18.99%	0.99%	0.35%	5.81%	550	2180	232
Moscow	Russia	0.30%	15.39%	48.84%	0.00%	16.35%	16.54%	2.57%	0.00%	0.00%	415	7243	202
Helena	USA	1.77%	38.23%	29.77%	0.09%	6.84%	19.04%	0.68%	0.11%	3.48%	538	4057	227
Östersund	Sweden	7.15%	1.01%	1.03%	0.00%	48.00%	37.76%	0.65%	0.01%	4.40%	40	18657	155
Yakutsk	Russia	0.30%	15.39%	48.84%	0.00%	16.35%	16.54%	2.57%	0.00%	0.00%	415	7243	202

2.5. Basis for Economic Assessment

The life cycle cost was used as a basis for economic assessment. According to the scheme proposed in [46], life cycle costs of buildings under an energy efficiency assessment framework may include initial, operation, repair, spare, downtime, loss, maintenance (corrective, preventive and predictive) and disposal costs. Based on available data, only initial (capital) costs of installations and operational costs of electricity and of water were considered.

For each configuration, the life cycle cost (LCC) for an interest rate of 10% and a lifetime of 10 years was calculated using the standard formula:

$$LCC = C_{op}\left(\frac{q^n - 1}{q^n \cdot i}\right) + C_{cap},\tag{4}$$

where:

- C_{op} is the operating cost.
- C_{cap} is the plant capital cost.
- i is the interest rate.
- n is the useful life of the plant.
- $q = 1 + i$.

A payback analysis was also later introduced to compare the economic feasibility between ABS and FC configurations. The payback period (PB) was evaluated using the following formula:

$$PB = \frac{C_{p,ABS} - C_{p,FC}}{C_{o,FC} - C_{o,ABS}},\tag{5}$$

where:

- $C_{p,ABS}$ is the plant cost of the ABS configuration.
- $C_{p,FC}$ is the plant cost of the FC configuration.
- $C_{o,FC}$ is the operating (energy and water) cost of FC alternatives.
- $C_{o,ABS}$ is the operating (energy and water) cost of ABS alternatives.

One should bear in mind that the simple payback time calculated according to Equation (5) allows a direct comparison between ABS and FC alternatives, but does not account for interest rates, which may lead to slightly different results with respect to LCC-based comparisons.

Capital cost estimates for absorption cooling systems, mechanical vapour compression cooling systems, dry coolers and cooling towers were obtained from the cost functions reported in [9], summarized in Table 3.

Table 3. Equipment cost functions used in this work.

Technology	Cost Function Structure (Y in €)
MVC Chiller	$Y = 20{,}000 + 112Q$ (Q cooling power in kW)
Absorption Chiller	$Y = 95{,}000 + 94Q$ (Q cooling power in kW)
Dry Cooler, Free Cooler	$Y = 8000\left(\frac{Q_d}{200}\right)^{0.7}$ (Qd dissipation capacity in kW)
Cooling Tower	$Y = 60{,}000\left(\frac{Q_d}{8000}\right)^{m}$ (Qd dissipation capacity in kW)

The operational expenses for different configurations were mainly determined by electricity consumption and—for CT configurations—by water consumption. It was not possible to retrieve electricity and industrial water prices all over the world. To obtain an approximate estimate for reasonable ranges, Western European values were derived from [9] as guide values, and are reported in Table 4.

Table 4. Prices of electricity and water used in this work.

	Unit	Min	Mean	Max
Electricity	Euro/kWh	0.073	0.112	0.175
Water	Euro/m^3	0.771	1.735	3.813

3. Results and Discussion

3.1. Average Total Cabin Cooling Load

Figure 5 shows the total cooling load, averaged over operating hours of one year, for the reference electric cabin in the climate zones of interest. On a yearly basis, heat transfer through the cabin envelope led to lower cooling loads, particularly in colder climates. However, the difference between the average cooling load in climate 7 (cold) and in climate 1A (hot) Was less than 3% of the total cooling load.

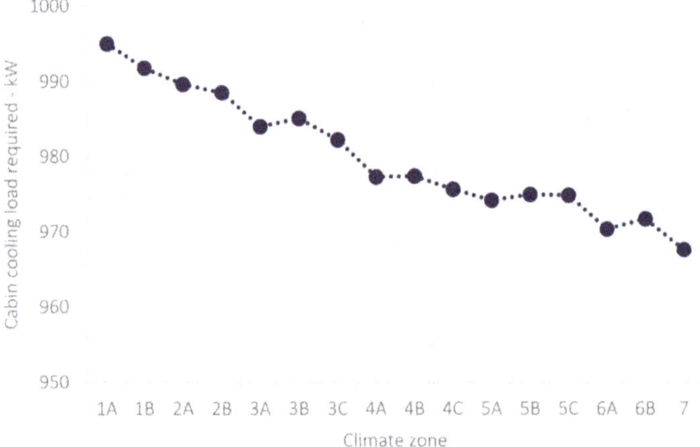

Figure 5. Simulated average cabin load over 16 climate zones.

3.2. Electric Energy Consumption

For each configuration, Figure 6 shows the annual electric energy consumption of the whole system (i.e., including cabin cooling as well as waste heat management from the WCD within the system boundaries). In Figures 6–9, the results are represented as histograms for each climate zone, in two rows (sub-figures) depending on the condenser used (with DC above and CT below). The bars labelled as BASE, FC, and ABS corresponding to the traditional MVC, free cooling, and absorption cooling configurations, respectively.

In Figure 6, the electricity demand by auxiliaries (AUX) is represented in green scale at the base of the bars, the demand for cabin refrigeration (REF) is represented by the intermediate bars in blue scale, while the top bars in yellow scale correspond to the electricity demand for heat dissipation from the WCD of the EAF. The relative reduction of electricity consumption in the ABS cooling mode compared with the baseline was on the order of 40% to 60%, and was more evident in CT configurations, which inherently require less electric power than DC to dissipate the same heat flows. In the FC configuration, which does not include waste heat recovery, the electricity consumption for heat dissipation at the WCD remained unchanged from baseline.

The reduction in electricity consumption for refrigeration was more pronounced in FC than in ABS configuration in several climatic regions, but almost exclusively when DCs were used. In fact, the overall balance resulted in a slight to remarkable advantage for ABS in all but the last climates in CT configurations, whereas in DC configurations FC outperformed the ABS configuration in climate zones 4C (mixed-marine), 5A (cool-humid), 5C (cool-marine) and 7. Interestingly enough, in continental dry climates such as in Dunhuang (5B) and Helena (6B), prolonged high-temperature periods in summer reduced the contribution of FC to electric energy saving over the year, making ABS cooling more attractive in terms of electricity demand.

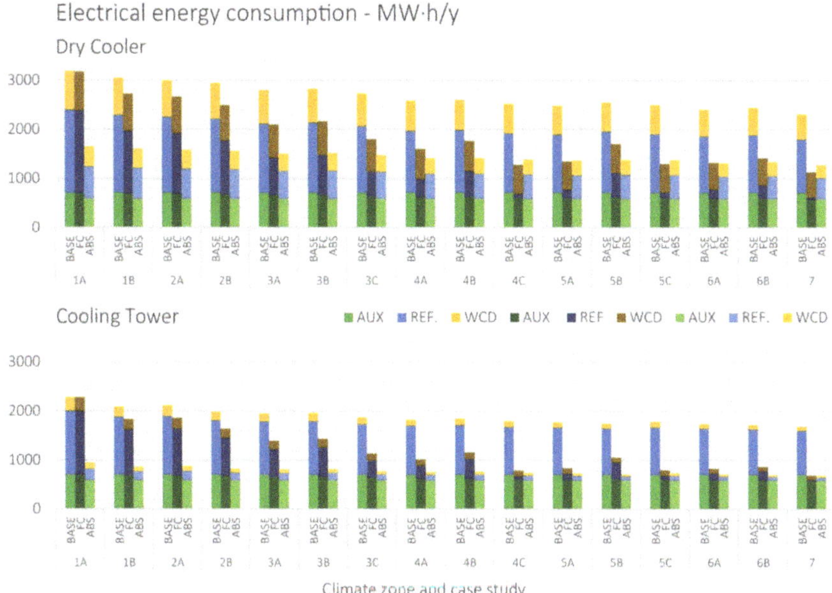

Figure 6. Electricity consumption based on climate zone and case study. AUX: electricity demand by auxiliaries; REF: demand for cabin refrigeration.

3.3. Direct and Indirect Water Consumption

Water consumption is illustrated in Figure 7. DC configurations imply only indirect water consumption, which was small and basically mirrored electricity consumption patterns, in a more or less pronounced way depending on the water intensity of the local electricity generation mix (see Table 2). In other words, this means that when DCs were used as heat dissipation systems, ABS configurations were mostly preferable to FC configurations also as to their impact on freshwater consumption.

For CT configurations, the direct water consumption (at the base of bars, in blue) was several orders of magnitude larger than the indirect consumption. Nevertheless, their overall balance was significantly affected by the indirect water footprint: for instance, although the direct water consumption of CTs evidently decreased in colder climates, the total water footprint of the baseline CT configuration in Östersund (climate region 7) equalled the baseline in Cairo (climate region 2B) due to the high indirect water footprint of electricity in Sweden.

For CT configurations, FCs were usually best performers in terms of direct freshwater footprint. In fact, the direct water consumption of FC configurations was always lower than that of corresponding ABS systems, except in very hot humid climates (1A). Indeed, the ABS configuration generally had a direct water consumption lower than or equal to the baseline, and its total (i.e., direct plus indirect) freshwater footprint was always lower than the baseline. However, FC configurations, in spite of their higher electricity demand, generally outperformed ABS cooling in terms of overall freshwater footprint, which was lower for the ABS configuration only in Singapore (1A) and New Delhi (1B). For New Delhi, this was mainly due to the high indirect water consumption associated with the national electricity mix, which is rich in water-intensive solid-fuel-based power generation (Table 2).

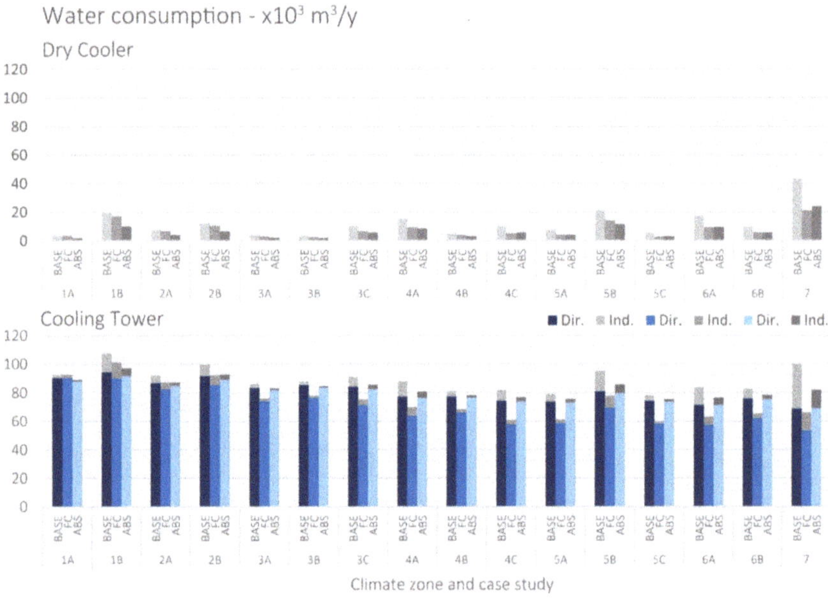

Figure 7. Direct and indirect water consumption based on climate zone and case study.

3.4. Direct and Indirect CO₂ Emissions

CO_2 footprints are compared in Figure 8. Only indirect CO_2 emissions associated with electricity consumption characterized the systems of interest. Hence, the CO_2 emission performance of configurations reflects electricity consumption both for DC and CT configurations. Indeed, the influence of the national energy mix on carbon emission was remarkable: for instance, baseline DC configurations achieved similar performance in Birmingham and Cairo, although their electricity consumption would be about 15% higher in the latter city (see Figure 6). Nevertheless, the climate dependence trend demonstrated for electricity (Figure 6) was basically conserved for CO_2 emissions (Figure 8): heat-recovery-based ABS cooling outperformed FC in all but the last CT configuration and in very hot to mixed climate zones, as well as in cool climates with hot summers, for DC configurations.

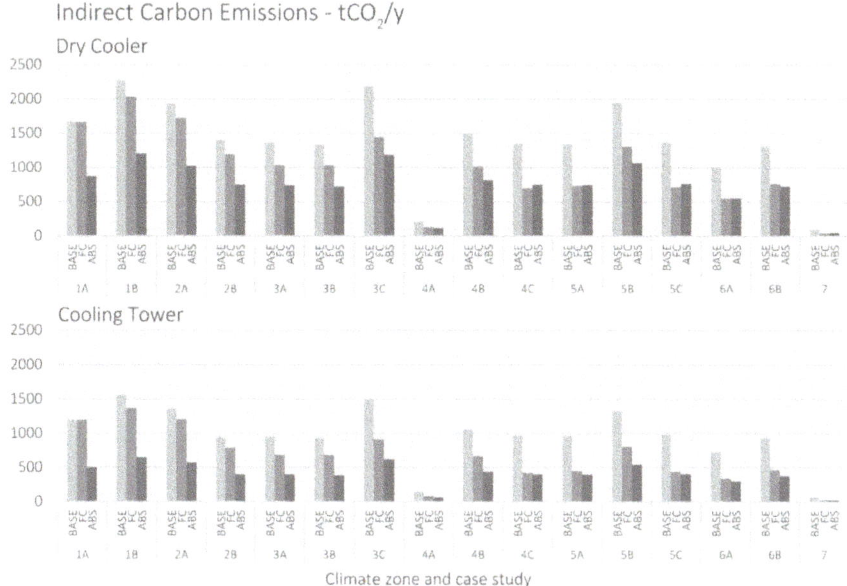

Figure 8. Indirect CO_2 emission based on climate zone and case study.

3.5. Primary Energy Consumption

The primary energy consumption trends represented in Figure 9 basically followed the CO_2 footprint trends. Comparing Figures 8 and 9, slight differences in the trends were observed in regions with higher shares of nuclear energy (e.g., Taipei—2A, Lyon—4A, Seoul—4B, Östersund—7), which has minimal GHG emissions but a high primary energy factor according to reference [45].

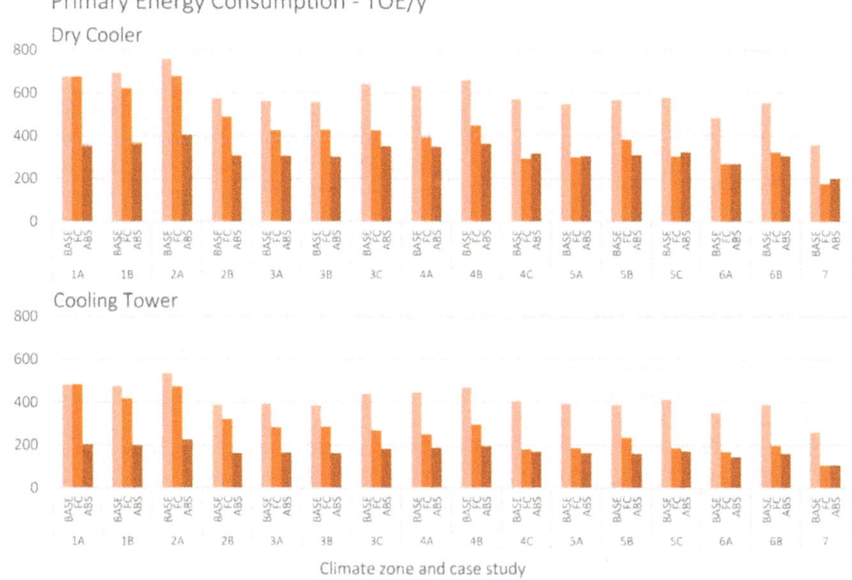

Figure 9. Primary energy (PE) consumption based on climate zone and case study.

3.6. Economic Performance

The LCCs of alternative cooling systems configurations are compared in Figure 10.

It is interesting to observe that at the resource cost conditions considered (average conditions, labelled as "Mean" in Table 4), the life cycle costs for configurations using CTs were slightly higher than corresponding configurations using DCs. In the long run, the lower capital costs and lower electricity cost of CT-based solutions did not compensate the additional economic impact of water consumption.

The proposed FC strategy requires few components and a simple regulation, which means that, compared with the baseline, additional investments are basically negligible. As a consequence, the life cycle costs of FC configurations were always lower than those of corresponding MVC configurations.

The proposed waste-heat to cooling ABS configuration requires significant investment but generates major savings in electricity costs and—for CT configurations—small savings in water costs in most climates. As a result, the life cycle costs of ABS configurations at average price conditions were always lower than those of corresponding MVC configurations, but in some climate regions (4C, 5A, 5C, 6A, 6B, 7 with either DCs or CTs, and also in 4A with CTs) they were higher than those of FC configurations, as shown in Figure 10.

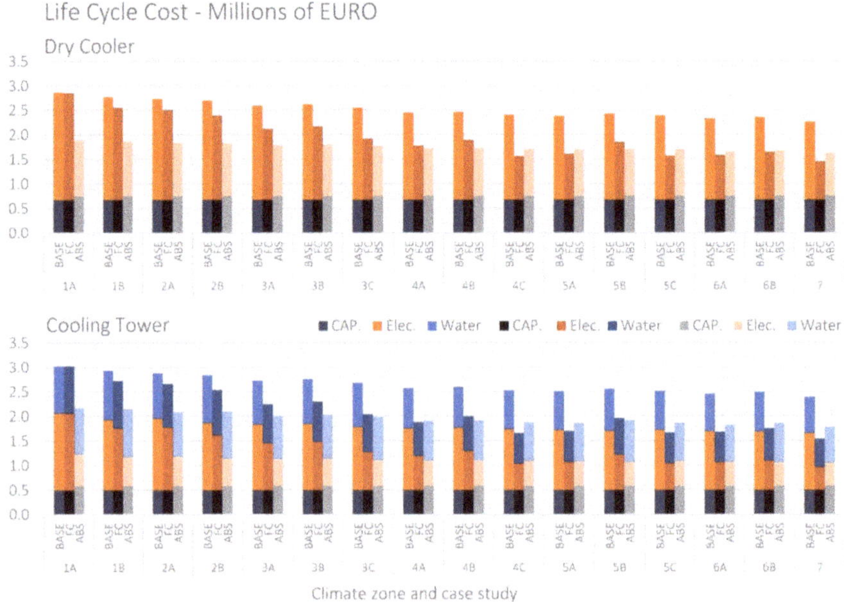

Figure 10. Life cycle cost (LCC) analysis based on climate zone and case study at mean electricity and water prices. CAP. is for plant capital costs, Elec. and Water are the life cycle operational expenses for electricity and water, respectively.

This is highlighted more generally in Figure 11, which shows the simple payback time (PBT) of ABS with respect to corresponding FC configurations according to Equation (5). Figure 11 also shows the dependence of PBT on climate zones, electricity prices and, for CT configurations, on water prices. In particular, Figure 11a shows the sensitivity of the economic performance of ABS with DC to the electricity price by presenting PBTs at minimum, mean and maximum levels of electricity price according to Table 4. Similarly, Figure 11b shows the sensitivity of PBTs in the ABS with CT configuration to the electricity price and Figure 11c shows the sensitivity of PBTs in the ABS with CT configuration to the water price (at the minimum, mean and maximum levels reported in Table 4). It can be observed that lower electricity prices resulted in smaller savings and consequently longer payback periods, which stretched even beyond the investment time horizon for CT configurations from climate zone 3C and subsequent. Similarly, high water prices in CT configurations increased operational expenses of ABS configurations, and hence led to longer payback times. Grey areas, corresponding to regions 4C, 5A, 5C, 6A and 7 with either DCs or CTs, and also in 6B with CTs) indicate the climate zones where the electricity consumption of ABS configuration was mostly higher than that of FC configuration: depending on price levels, the payback times of ABS configurations in these cases would either be negative (i.e., ABS never pays off) or at least longer than the system's service life.

It can be observed that:

- ABS cooling was more cost effective than FC in most climates and paid off in less than three years even at worst (i.e., lowest) electricity price conditions in very hot to warm dry or humid climate zones, both in DC and CT configurations;
- Where DCs are used (e.g., due to general water scarcity), ABS cooling may be a more rewarding option than FC even in mixed to cool dry climate zones (4B and 5B). Similar paybacks were also achieved with DC in warm marine climate zones (3C), however it should be noted that if industries are placed directly by the sea-coast, other resource-efficient heat rejection options may be used (e.g., once-through cooling) which are beyond the scope of the present research;
- ABS CT configurations, featuring lower electricity consumption and absolute electric energy savings than DC, had slightly longer PBTs, which were nevertheless satisfactory (i.e., lower than three years) in very hot to warm dry climates, and in unfavourable economic conditions (low electricity prices or high water prices).
- The economic performance of ABS systems with CTs was more sensitive to electricity price in hot climates, and to water prices in mixed to cool dry climates where FC enables substantial water savings compared with ABS cooling (see Figure 7).

(a)

Figure 11. *Cont.*

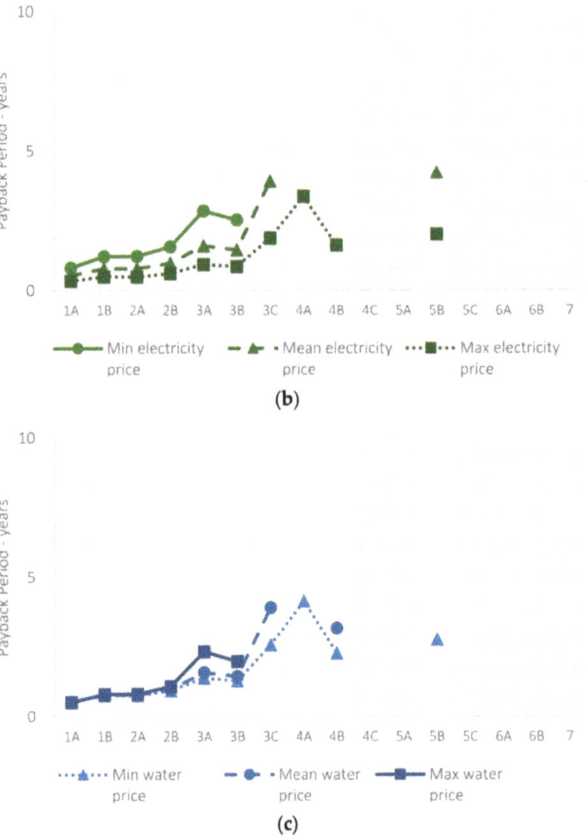

Figure 11. Payback times of ABS cooling solutions with respect to FC solutions depending on electricity and water prices (**a**) FC with DC, sensitivity to electricity price; (**b**) FC with CT, sensitivity to electricity price; (**c**) FC with CT, sensitivity to water price.

4. Conclusions

In this study, the energy, water, CO_2 and primary energy consumption of different configurations of air conditioning systems for electric cabins were analysed considering 16 ASHRAE climate zones worldwide. The case at hand refers to electric cabins of EAF steelmaking sites, but the methodology and general results could be easily extended to similar installations in process industries having considerable low-grade waste heat flows.

It was confirmed that the proposed waste heat utilization system for absorption cooling allowed substantial energy savings and an overall favourable water–energy–GHG balance in all climate zones compared with traditional mechanical vapour compression air cooling systems for electric cabins. Compared with the simple airside free cooling configuration proposed here, absorption cooling was also the better option as to electric energy consumption in nearly all climate zones (15 out of 16 climate zones for systems with cooling towers and 12 out of 16 climate zones for systems with dry coolers as heat rejection units). A similar trend was observed for GHG emissions and primary energy demand, which had more or less pronounced differences depending on the local electricity generation mix of the analysed locations. In some cases, such indicators were shown to be more heavily affected by the country's energy mix than by climatic conditions.

However, it was shown that in configurations using cooling towers, free cooling led to significantly lower direct water consumption in nearly all climate regions, and to a lower overall water footprint, particularly in mixed to very cold climates. In climate regions 4B to 7, free cooling was also more cost effective than waste-heat recovery for the examined ranges of economic parameters. In those regions, free cooling also allowed substantial energy savings, though lower than those of corresponding absorption cooling solutions.

Hence, in general terms, our results indicate that waste heat recovery for electric cabin cooling is an energy efficient solution which is also water efficient when dry cooling systems are used. In systems with cooling towers, it generally decreases the system's water footprint compared with baseline configurations, but not with free cooling alternatives. Under the technical and market conditions examined in this paper, absorption cooling was clearly the least cost option for electric cabin cooling in warm to very hot climate zones, but where CTs are used it is the most water-efficient option only in very hot dry climates.

Due to the better performance of absorption cooling in terms of energy and carbon footprint, even in cold climates and with a low-carbon national energy generation mix, carbon- or energy-efficient incentives could make it economically feasible also in further climate regions (e.g., in mixed climate zones), even for systems using cooling towers and at high water prices. In that case, industrial designers and energy managers might prefer absorption cooling solutions because of their energy efficiency even where, compared with FC, they are suboptimal in terms of water consumption.

It is thus recommended that, particularly in intermediate climate conditions, decision makers accurately evaluate the water footprint of their energy efficiency projects; the approach proposed in the current analysis could support them in this task. It is also desirable that policy makers who design incentives supporting industrial energy efficiency or GHG reduction combine them with constraints or incentives for water consumption reduction, taking a nexus approach.

Author Contributions: Conceptualization: D.C. and O.S.; Methodology: collectively developed by all authors; Investigation: M.S. and A.Z.; Software and Visualization: M.S. and A.Z.; Supervision: D.C., O.S. and A.D.A.; Writing—original draft: D.C. and M.S.; Writing—review and editing: O.S., A.Z. and A.D.A.

Funding: This research received no external funding.

Conflicts of Interest: The authors declare no conflict of interest.

Nomenclature

ABS(C)	Absorption Chiller
AUX	Electricity demand by auxiliaries
BF	Blast Furnace
BOF	Basic Oxygen Furnace
$CO2_f$	Total CO_2 footprint—tons
$CO2_{ind}$	Total CO_2 indirect emissions—tons
$C_{CO_2,el}$	Carbon footprint coefficients for electricity consumption (tCO_2/GWh)
C_{inv}	Plant investment cost (EURO)
C_{op}	Operating cost (EURO/year)
$C_{o,ABS}$	ABS configuration operating cost (EURO/year)
$C_{o,FC}$	FC configuration operating cost (EURO/year)
$C_{PED,el}$	Site-to-source energy conversion factors (TOE/GWh)
$C_{p,ABS}$	Plant cost of ABS configuration (EURO)
$C_{p,FC}$	Plant cost of FC configuration (EURO)
$C_{W,el}$	Indirect water consumption rate (m^3/GWh)
CT	Cooling Tower
COP	Coefficient of Performance

DC	Dry Coolers
EAF	Electric Arc Furnace
E_{el}	Total electricity demand (GWh)
EER	Energy Efficiency Ratio
ETS	Emission Trading Schemes
EU	European Union
FC	Free Cooling
GHG	Greenhouse Gas
HVAC	Heating, Ventilation and Air Conditioning
i	Interest rate (%)
IWH	Industrial Waste Heat
k	Multiplicative coefficient for water losses due to bleed off and drift—dimensionless
MVC(C)	Mechanical Vapour Compression Chiller
n	Life of the plant (years)
ORC	Organic Rankine Cycle
PED	Primary Energy Demand (consumption) (TOE)
PB	Payback Period (years)
PBT	Payback Time
q	Defined as $1 + i$
REF	Refrigeration
TOE	Ton (of) Oil Equivalent
WCD	Water Cooled Duct
WEN	Water Energy Nexus
W_d	Direct water use (m^3)
W_{ev}	Evaporated water (m^3)
W_f	Total water footprint (m^3)
W_{ind}	Indirect water use (m^3)

References

1. He, K.; Wang, L. A review of energy use and energy-efficient technologies for the iron and steel industry. *Renew. Sustain. Energy Rev.* **2017**, *70*, 1022–1039. [CrossRef]
2. IEA. *World Energy Outlook 2010*; International Energy Agency: Paris, France, 2010; Available online: https://webstore.iea.org/world-energy-outlook-2010 (accessed on 17 June 2019).
3. European Union. European Commission, EU-ETS Handbook 2015. Available online: https://ec.europa.eu/clima/sites/clima/files/docs/ets_handbook_en.pdf (accessed on 17 June 2019).
4. Huisingh, D.; Zhang, Z.; Moore, J.C.; Qiao, Q.; Li, Q. Recent advances in carbon emissions reduction: Policies, technologies, monitoring, assessment and modeling. *J. Clean. Prod.* **2015**, *103*, 1–12. [CrossRef]
5. Keplinger, T.; Haider, M.; Steinparzer, T.; Patrejko, A.; Trunner, P.; Haselgrübler, M. Dynamic simulation of an electric arc furnace waste heat recovery system for steam production. *Appl. Therm. Eng.* **2018**, *135*, 188–196. [CrossRef]
6. Schnoor, J. Water-energy nexus. *Environ. Sci. Technol.* **2011**, *45*, 5065. [CrossRef] [PubMed]
7. World Steel Association. *Water Management in the Steel Industry 2015*; Position Paper; WSA AISBL: Brussels, Belgium, 2015; ISBN 978-2-930069-81-4. Available online: https://www.worldsteel.org/en/dam/jcr:f7594c5f-9250-4eb3-aa10-48cba3e3b213/Water+Management+Position+Paper+2015.pdf (accessed on 17 June 2019).
8. Wang, C.; Zheng, X.; Cai, W.; Gao, X.; Berrill, P. Unexpected water impacts of energy-saving measures in the iron and steel sector: Tradeoffs or synergies? *Appl. Energy* **2017**, *205*, 1119–1127. [CrossRef]
9. Chinese, D.; Santin, M.; Saro, O. Water-energy and GHG nexus assessment of alternative heat recovery options in industry: A case study on electric steelmaking in Europe. *Energy* **2017**, *141*, 2670–2687. [CrossRef]
10. Gao, C.; Wang, D.; Dong, H.; Cai, J.; Zhu, W.; Du, T. Optimization and evaluation of steel industry's water-use system. *J. Clean. Prod.* **2011**, *19*, 64–69. [CrossRef]

11. Gu, Y.; Xu, J.; Keller, A.A.; Yuan, D.; Li, Y.; Zhang, B.; Wenig, Q.; Zhang, X.; Deng, P.; Wang, H.; et al. Calculation of water footprint of the iron and steel industry: A case study in Eastern China. *J. Clean. Prod.* **2015**, *92*, 274–281. [CrossRef]

12. Moya, J.A.; Pardo, N. The potential for improvements in energy efficiency and CO_2 emissions in the EU27 iron and steel industry under different payback periods. *J. Clean. Prod.* **2013**, *52*, 71–83. [CrossRef]

13. Johansson, M.T.; Söderström, M. Options for the Swedish steel industry energy efficiency measures and fuel conversion. *Energy* **2011**, *36*, 191–198. [CrossRef]

14. Lin, Y.P.; Wang, W.H.; Pan, S.Y.; Ho, C.C.; Hou, C.J.; Chiang, P.C. Environmental impacts and benefits of Organic Rankine Cycle power generation technology and wood pellet fuel exemplified by electric arc furnace steel industry. *Appl. Energy* **2016**, *183*, 369–379. [CrossRef]

15. Lee, B.; Sohn, I. Review of innovative energy savings technology for the electric arc furnace. *JOM* **2014**, *66*, 1581–1594. [CrossRef]

16. Xu, Z.; Wang, R. Absorption refrigeration cycles: Categorized based on the cycle construction. *Int. J. Refrig.* **2016**, *62*, 114–136. [CrossRef]

17. Sun, J.; Fu, L.; Zhang, S. A review of working fluids of absorption cycles. *Renew. Sustain. Energy Rev.* **2012**, *16*, 1899–1906. [CrossRef]

18. Banu, P.A.; Sudharsan, N. Review of water based vapour absorption cooling systems using thermodynamic analysis. *Renew. Sustain. Energy Rev.* **2018**, *82*, 3750–3761. [CrossRef]

19. Shirazi, A.; Taylor, R.A.; Morrison, G.L.; White, S.D. Solar-powered absorption chillers: A comprehensive and critical review. *Energy Convers. Manag.* **2018**, *171*, 59–81. [CrossRef]

20. Wang, J.; Yan, R.; Wang, Z.; Zhang, X.; Shi, G. Thermal Performance Analysis of an Absorption Cooling System Based on Parabolic Trough Solar Collectors. *Energies* **2018**, *11*, 2679. [CrossRef]

21. Eicker, U. *Energy Efficient Buildings with Solar and Geothermal Resources*; John Wiley & Sons: Hoboken, NJ, USA, 2014.

22. Viklund, S.B.; Johansson, M.T. Technologies for utilization of industrial excess heat: Potentials for energy recovery and CO_2 emission reduction. *Energy Convers Manag.* **2014**, *77*, 369–379. [CrossRef]

23. Xia, L.; Liu, R.; Zeng, Y.; Zhou, P.; Liu, J.; Cao, X.; Xiang, S. A review of low-temperature heat recovery technologies for industry processes. *Chin. J. Chem. Eng* **2018**. In Press, Corrected Proof. [CrossRef]

24. Liew, P.Y.; Walmsley, T.G.; Alwi, S.R.W.; Manan, Z.A.; Klemeš, J.J.; Varbanov, P.S. Integrating district cooling systems in locally integrated energy sectors through total site heat integration. *Appl. Energy* **2016**, *184*, 1350–1363. [CrossRef]

25. Leong, Y.T.; Chan, W.M.; Ho, Y.K.; Isma, A.I.M.I.A.; Chew, I.M.L. Discovering the potential of absorption refrigeration system through industrial symbiotic waste heat recovery network. *Chem. Eng. Trans.* **2017**, *61*, 1633–1638.

26. Brückner, S.; Liu, S.; Miró, L.; Radspieler, M.; Cabeza, L.F.; Lävemann, E. Industrial waste heat recovery technologies: An economic analysis of heat transformation technologies. *Appl. Energy* **2015**, *151*, 157–167. [CrossRef]

27. Cola, F.; Romagnoli, A.; Hey, J. An evaluation of the technologies for heat recovery to meet onsite cooling demands. *Energy Convers. Manag.* **2016**, *121*, 174–185. [CrossRef]

28. Haywood, A.; Sherbeck, J.; Phelan, P.; Varsamopoulos, G.; Gupta, S.K. Thermodynamic feasibility of harvesting data centre waste heat to drive an absorption chiller. *Energy Convers. Manag.* **2012**, *58*, 26–34. [CrossRef]

29. Ebrahimi, K.; Jones, G.F.; Fleischer, A.S. A review of data centre cooling technology, operating conditions and the corresponding low-grade waste heat recovery opportunities. *Renew. Sustain Energy Rev.* **2014**, *31*, 622–638. [CrossRef]

30. Oró, E.; Depoorter, V.; Pflugradt, N.; Salom, J. Overview of direct air free cooling and thermal energy storage potential energy savings in data centres. *Appl. Therm. Eng.* **2015**, *85*, 100–110. [CrossRef]

31. Daraghmeh, H.M.; Wang, C.C. A review of current status of free cooling in datacentres. *Appl. Therm. Eng.* **2017**, *114*, 1224–1239. [CrossRef]

32. Agrawal, A.; Khichar, M.; Jain, S. Transient simulation of wet cooling strategies for a data centre in worldwide climate zones. *Energy Build.* **2016**, *127*, 352–359. [CrossRef]

33. Thermal Energy System Specialists LLC—TRNSYS 17 Documentation, Mathematical Reference. Available online: http://web.mit.edu/parmstr/Public/TRNSYS/04-MathematicalReference.pdf (accessed on 17 June 2019).

34. ASHRAE. *ASHRAE/IESNA 90.1 Standard-2007-Energy Standard for Buildings Except Low-Rise Residential Buildings*; American Society of Heating Refrigerating and Air-conditioning Engineers: New York, NY, USA, 2007.

35. Pansera, G.; Griffini, N. Dedusting plants for electric arc furnaces. *Millenn. Steel.* **2016**, 85–89. Available online: http://millennium-steel.com/wp-content/uploads/articles/pdf/2005/pp85-89%20MS05.pdf (accessed on 17 June 2019).

36. Kühn, R.; Geck, H.G.; Schwerdtfeger, K. Continuous off-gas measurement and energy balance in electric arc steelmaking. *ISIJ Int.* **2005**, *45*, 1587–1596. [CrossRef]

37. Johnson Controls. *York® Commercial & Industrial HVAC 2016*; Johnson Controls: Milwaukee, WI, USA, 2017; Available online: http://www.clima-trade.com/Download/be_york_chillers_and_heatpumps_en_2016.pdf (accessed on 17 June 2019).

38. Mombeni, A.G.; Hajidavalloo, E.; Behbahani-Nejad, M. Transient simulation of conjugate heat transfer in the roof cooling panel of an electric arc furnace. *Appl. Therm. Eng.* **2016**, *98*, 80–87. [CrossRef]

39. Johnson Controls. *York® YIA Absorption Chiller Engineering Guide*; Johnson Controls: Milwaukee, WI, USA, 2017; Available online: https://www.johnsoncontrols.com/-/media/jci/be/united-states/hvac-equipment/chillers/be_engguide_yia_singleeffect-absorption-chillers-steam-and-hot-water-chillers.pdf (accessed on 17 June 2019).

40. LG Electronics. LG HVAC Solution Absorption Chiller. 2018. Available online: https://www.lg.com/global/business/download/resources/CT00022379/CT00022379_28284.pdf (accessed on 17 June 2019).

41. LU-VE/AIA. Dry Coolers and Condensers for Industrial Applications. 2017. Available online: https://manuals.luve.it/Industrial%20Applications/files/assets/common/downloads/Industrial%20Applications.pdf (accessed on 17 June 2019).

42. YWCT. YWCT Cooling Tower Catalogue. 2010. Available online: http://www.customcoolingtowers.com/_Uploads/dbsAttachedFiles/YWCT_Catalog_2010_2.pdf (accessed on 17 June 2019).

43. Energy Plus Climate Database. 2018. Available online: https://energyplus.net/ (accessed on 17 June 2019).

44. Chhipi-Shrestha, G.; Kaur, M.; Hewage, K.; Sadiq, R. Optimizing residential density based on water–energy–carbon nexus using UTilités Additives (UTA) method. *Clean Technol. Environ. Policy* **2018**, *20*, 855. [CrossRef]

45. Waas, T.; Hugé, J.; Block, T.; Wright, T.; Benitez-Capistros, F.; Verbruggen, A. Sustainability assessment and indicators: Tools in a decision-making strategy for sustainable development. *Sustainability* **2014**, *6*, 5512–5534. [CrossRef]

46. Danish, M.S.S.; Senjyu, T.; Ibrahimi, A.M.; Ahmadi, M.; Howlader, A.M. A managed framework for energy-efficient building. *J. Build. Eng.* **2019**, *21*, 120–128. [CrossRef]

47. REA, Inc. Cooling Tower Make-up Water Flow Calculation. Available online: http://www.reahvac.com/tools/cooling-tower-make-water-flow-calculation/ (accessed on 17 September 2019).

48. World Bank Database. World Development Indicators: Electricity Production, Sources, and Access. Available online: http://www.tsp-data-portal.org/Breakdown-of-Electricity-Generation-by-Energy-Source#tspQvChart (accessed on 18 June 2019).

49. Pandey, D.; Agrawal, M.; Pandey, J.S. Carbon footprint: Current methods of estimation. *Environ. Monit. Assess.* **2011**, *178*, 135–160. [CrossRef] [PubMed]

50. Meunier, F. Co- and tri-generation contribution to climate change control. *Appl. Therm. Eng.* **2002**, *22*, 703–718. [CrossRef]

51. Wilby, M.R.; Gonzalez, A.B.R.; Díaz, J.J.V. Empirical and dynamic primary energy factors. *Energy* **2014**, *73*, 771–779. [CrossRef]

Article

Deformation Effect on Water Transport through Nanotubes

Ferlin Robinson, Majid Shahbabaei and Daejoong Kim *

Department of Mechanical Engineering, Sogang University, 35 Baekbeom-ro, Mapo-gu, Seoul 121-742, Korea;
ferlin@sogang.ac.kr (F.R.); m.shahbabaei@gmail.com (M.S.)
* Correspondence: daejoong@sogang.ac.kr

Received: 31 October 2019; Accepted: 18 November 2019; Published: 21 November 2019

Abstract: In this study, we used non-equilibrium molecular dynamics to study the transport of water through deformed (6,6) Carbon Nanotubes (CNTs) and Boron Nitride Nanotubes (BNNTs). The results were then compared with that of the perfect nanotubes. The main aim of this study was to get a better insight into the deformation effect on water transport through nanotubes rather than directly comparing the CNTs and BNNTs. As the diameters of both types of nanotubes differ from each other for the same chiral value, they are not directly comparable. We carried out our study on deformations such as screw distortion, XY-distortion, and Z-distortion. XY-distortion of value 2 shows a change from single-file water transport to near-Fickian diffusion. The XY-distortions of higher value shows a notable negative effect on water transport when their distortion values get larger. These suggest that the degree of deformation plays a crucial role in water transport through deformed nanotubes. The Z-distortion of 2 showed discontinuous single-file chain formation inside the nanotubes. Similar phenomena are observed in both nanotubes, irrespective of their type, while the magnitudes of their effects vary.

Keywords: non-equilibrium molecular dynamics; deformed carbon nanotubes; deformed boron nitride nanotubes; water transport; diffusion; Z-distortion; XY-distortion; screw distortion

---

1. Introduction

Nanotubes (NTs) are prominent structures in many applications. From water desalination to microelectronics, their areas of application are wide. Their application areas are even getting broad every day, as they hold a promising future due to their distinctive properties. The discovery of carbon nanotubes (CNTs) by Sumio Iijima [1] has revolutionized the nano world. Since its discovery, the field of nanotechnology has achieved greater heights. Rubio et al. [2] predicted boron nitride nanotubes (BNNTs) in 1994, and they were experimentally discovered in 1995, by Chopra et al. [3]. In 2001, Hummer et al. [4] found that water can pass through CNTs spontaneously, using molecular dynamics simulation. Since then, many studies were carried out, using perfect CNTs and BNNTs for water transport and desalination. But the studies carried out on the deformed nanotubes are limited. Different types of deformation occur in nanotubes. They could be twisted, compressed, elongated, or bent during application and manufacturing. They could also have defects such as Stone–Wales, point vacancies, interstitials, etc., during manufacturing. He et al. [5] studied the effects of deformation degree and about the location of deformation in the carbon nanotubes on water transport. Feng et al. [6] showed that the transport diffusion of helium gas through deformed carbon nanotubes with screw deformation did not have any effect, while the XY-distortion and Z-distortion showed that the effect on transport diffusion is significant with an increase in temperature and distortions values. Even though there were earlier studies carried out on the water transport phenomena through deformed carbon nanotubes, water transport through nanotubes that have a twist or XY-distortion and Z-distortion have not been studied extensively yet. Therefore, in this work, we studied the effect of deformations such as screw distortion, XY-distortion, and Z-distortion on water transport through long CNTs and BNNTs.

The effect of deformation on water transport through nanotubes is also of great importance to many biological, ion-selective channels. The ionic conduction through single-walled carbon nanotubes can directly be compared to them. Ion-channels are formed from proteins [7]. CNT-based nanodevices with controllable functions are widely used to mimic transmembrane channels. These transmembrane channels can be used as substitutes for certain channel proteins. Model membranes are needed to examine their application as transmembrane channels. Two methods are widely used to simulate this bio-membrane system [8]. The first method is to embed the CNT as channel in-between graphite sheets as the membrane. The atoms are deleted at desired locations, in order to accommodate the tube [9–11]. The second method is the CNT-bundle system. In this method, packed CNTs serve both as channels and the membrane to separate the reservoirs [12,13]. In this study, we used the first method. This study about the effect of deformation can be of greater use for designing transmembrane channels.

2. Simulation Details

2.1. Computational Domain and Structures

Figure 1 shows the schematic of the molecular dynamics setup used in this study for CNT and BNNT. In this study, we used CNTs and BNNTs with chirality (6,6) and length 100 Å in all simulations. As the Z-distortion name suggests, the nanotube deformation is along the Z-direction; hence, a nanotube of length 100 Å gets distorted to 90, 110, and 120 Å for distortion values of 0.9, 1.1, and 1.2, respectively. To compare the effect of different deformations on water transport, at least one physical property has to be kept constrained. In this study, the length was kept constrained; hence, the Z-distortion nanotubes were modeled considerably long, and then they were truncated to have the length ~100 Å.

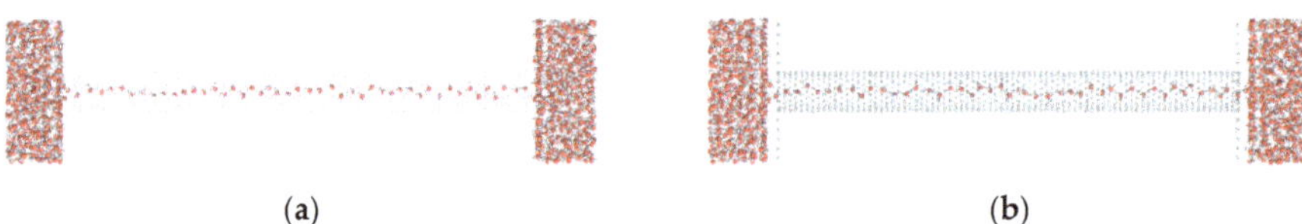

(a)   (b)

Figure 1. Schematic of molecular dynamics simulation setup: (**a**) boron nitride nanotube (BNNT) and (**b**) carbon nanotube (CNT).

The geometry of the perfect nanotube is shown in Figure 2a. The Z-distortion is induced in the perfect nanotube either by stretching or compressing along Z-axis, which increases or decreases the length of the nanotube, respectively. The stretch factor is given by Z'/Z_0, where Z' is the total length after the perfect nanotube is distorted in the Z-direction [6]. The Z-distorted nanotube is shown in Figure 2b. The nanotube with the twist is shown in Figure 2c. The screw factors for the twisted nanotubes are given as β/Z_0, where β is the twist angle and Z_0 is the length of the perfect nanotube. When an equal amount of force is applied in the positive and negative X or Y direction, XY-distorted carbon nanotubes can be obtained, as shown in Figure 2d. The XY-distorted nanotubes have an elliptical cross-section with ellipse factor $\Delta X/\Delta Y$.

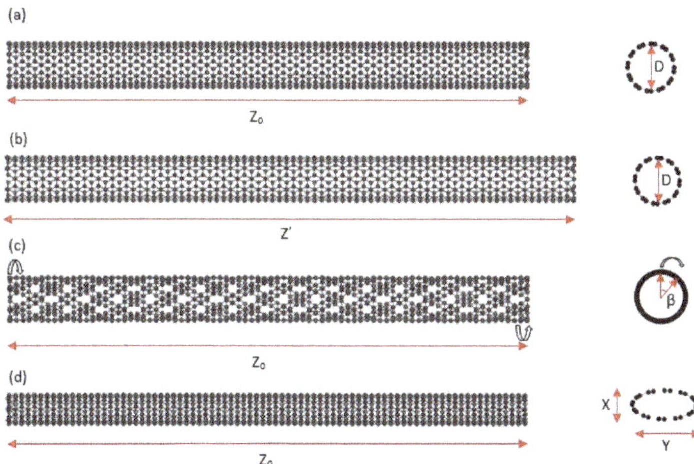

Figure 2. Perfect nanotube and different types of deformed nanotubes: (**a**) perfect nanotube; (**b**) Z-distortion nanotube (stretch factor: Z'/Z_0); (**c**) twisted nanotube (screw factor: β/Z_0); and (**d**) XY-distortion nanotube (ellipse factor: e = $\Delta X/\Delta Y$).

The perfect (6,6) CNT has a radius of 4.068 Å and bond length of 1.421 Å, whereas a perfect (6,6) BNNT has a radius of 4.211 Å and bond length of 1.47 Å. The description of the dimensions of the nanotubes is given in Table 1.

Table 1. Dimensions of different nanotubes used in this study. Z-distortion nanotube (stretch factor: $Z'/Z0$), twisted nanotube (screw factor: β/Z_0), and XY-distortion nanotube (ellipse factor: e = $\Delta X/\Delta Y$).

Nanotube Type	Distortion Type	Major Axis Value (Å)	Minor Axis Value (Å)	Radius (Å)
CNT	Perfect NT	-	-	4.068
CNT	Screw Distortion 15	-	-	4.068
CNT	Screw Distortion 30	-	-	4.068
CNT	Screw Distortion 45	-	-	4.068
CNT	XY-distortion 2	10.560	6.277	-
CNT	XY-distortion 4	11.928	5.557	-
CNT	XY-distortion 6	12.331	5.376	-
CNT	Z-distortion 0.9	-	-	4.068
CNT	Z-distortion 1.1	-	-	4.068
CNT	Z-distortion 1.2	-	-	4.068
BNNT	Perfect NT	-	-	4.211
BNNT	Screw Distortion 15	-	-	4.211
BNNT	Screw Distortion 30	-	-	4.211
BNNT	Screw Distortion 45	-	-	4.211
BNNT	XY-distortion 2	11.217	6.494	-
BNNT	XY-distortion 4	12.599	5.749	-
BNNT	XY-distortion 6	13.008	5.561	-
BNNT	Z-distortion 0.9	-	-	4.211
BNNT	Z-distortion 1.1	-	-	4.211
BNNT	Z-distortion 1.2	-	-	4.211

2.2. Computational Methods

All molecular dynamics simulations were carried out by using the open-source software DL POLY 4.08 [14,15]. For visualization, VMD 1.9.4a12 software [16] was used. All simulations were carried out by using the non-equilibrium molecular dynamics (NEMD) method to derive the motion of the molecules. LJ-potential is used to describe the interactions between the water molecules and the interaction between the CNT and water molecules. The LJ-potential is given by the following equation:

$$U(r_{ij}) = 4\epsilon[(\sigma/r_{ij})^{12} - (\sigma/r_{ij})^6]$$

where r_{ij}, σ, and ϵ are the interatomic bond vector, the balance distance where the interaction of particles is zero, and the depth of the potential well, respectively.

In this study, to model the flow, an external force field method is used. In this method, a constant force f is added to all water molecules along +Z direction to mimic the pressure-driven flow, whereas the pressure difference between the two sides of a membrane is as follows:

$$\Delta P = nf/A$$

where f is the applied force on each water molecule, n is the number of water molecules, and A is the membrane area. This is one of the widely accepted methods in molecular dynamics (MD) to simulate the pressure-driven flow [17]. Force is applied to the molecules, using the external force field known as the gravitational field in the DL POLY software, where the force is given by the following equation:

$$\underline{F} = m\underline{G}$$

where $\underline{G}$ is the gravitational field which is given as input, $\underline{F}$ is the force, and m is the mass of the water molecule.

All simulations were carried out with a gravitational field of 0.0185G applied to the molecules. The SPC/E water model used in this simulation has a contact angle $\theta = 95.3°$ [18]. The LJ cross-interaction between the water molecules and carbon atoms are $\sigma_{C-O} = 0.319$ nm and $\epsilon_{C-O} = 0.392$ kJ/mol [18], which were determined by using the Lorentz-Berthelot mixing rules [19]. For water–water interactions, the default value of LJ parameter of SPC/E water model $\sigma_{O-O} = 0.3169$ nm and $\epsilon_{O-O} = 0.6498$ kJ/mol is used [20]. The LJ parameters of $\sigma_{B-O} = 0.331$ nm, $\epsilon_{B-O} = 0.5079$ kJ/mol, $\sigma_{N-O} = 0.326$ nm, and $\epsilon_{N-O} = 0.6276$ kJ/mol were used for the boron nitride interactions with water [21]. The bond lengths and angle degrees of the water molecule are constrained by SHAKE algorithm [22]. The canonical ensemble NVT is used for all the simulations, to update the velocity and position, along with Noosé-Hoover thermostat coupling, to maintain a constant temperature. A cut-off distance of 10 Å is used for LJ interactions.

3. Results and Analyses

A few key features were observed in the water transport phenomena through deformed nanotubes, including a change of single-file water chain transformation to two single-file water chains forming next to each other in the XY-distortion of 2 nanotubes, resembling a near-Fickian diffusion; formation of discontinuous single-file chain water transport in Z-distortion of 1.2 nanotubes; and a smattering water transport through XY-distortion of 4 and 6 nanotubes. Figure 3 shows the water transport through a perfect BNNT nanotube and the XY-distortion 2 BNNT nanotube.

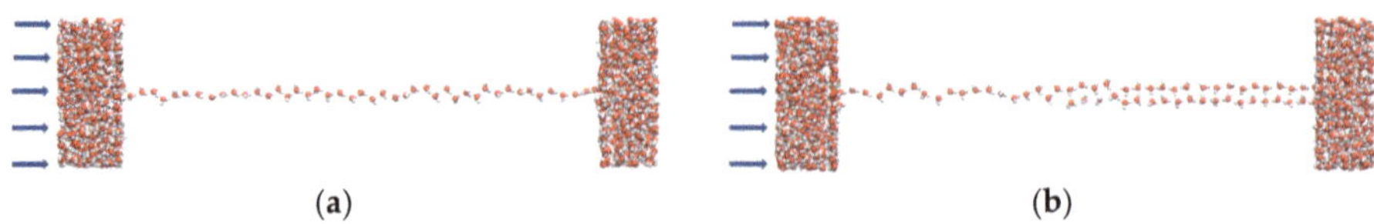

(a)   (b)

Figure 3. Visualization of water chain formed in-between two reservoirs: (a) perfect BNNT (b) BNNT with XY-distortion value 2.

3.1. Free Energy of Occupancy Fluctuations and Water Occupancy

Figure 4 shows the free energy of occupancy fluctuations as a function of the number of water molecules inside the nanotube for the CNT and BNNT with XY-distortion. The free energy of the occupancy fluctuations is calculated as follows:

$$\beta F(N) = -\ln p(N)$$

where $p(N)$ is the probability of finding exactly 'N' water molecules inside the nanotube, $F(N)$ is the free energy and $\beta = (k_B T)^{-1}$, k_B is the Boltzmann constant, and T is the temperature [4,23].

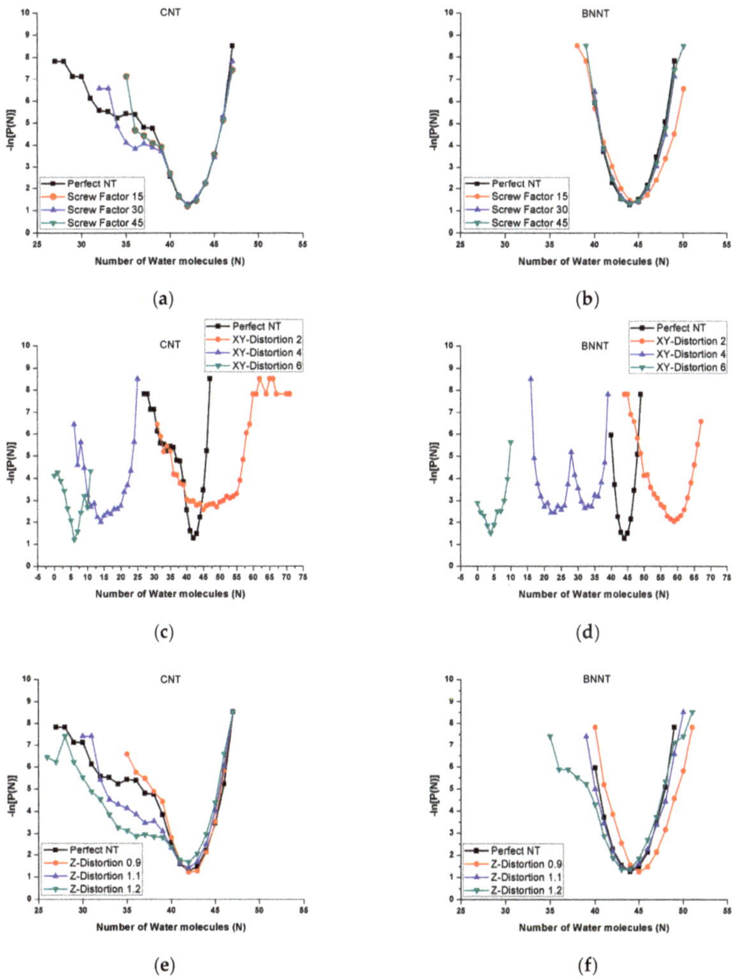

Figure 4. Free energy of occupancy fluctuation as a function of number of water molecules inside and different types of deformed nanotubes compared with perfect nanotubes: (**a**) carbon nanotube (CNT) with screw distortions; (**b**) boron nitride nanotube (BNNT) with screw distortions; (**c**) carbon nanotube (CNT) with XY-distortion; (**d**) boron nitride nanotube (BNNT) with XY-distortion; (**e**) carbon nanotube (CNT) with Z-distortion nanotube; and (**f**) boron nitride nanotube (BNNT) with Z-distortion.

The plot shows that the number of water molecules inside the perfect CNT ranges from 26 to 47. The most probable number of molecules is 42. The number of molecules inside the perfect BNNT ranges from 40 to 48. The most probable number is 44. It can be seen that the XY-distortion of 2 nanotubes accommodates the maximum number of water molecules when compared to others. The XY-distorted nanotubes of higher value show a significantly lower number of molecules in both the carbon and boron nitride nanotubes. The reason for the XY-distortion nanotubes to accommodate the maximum number of water molecules is that, when a nanotube is distorted in the XY-direction with the distortion value of 2, it has sufficient space to accommodate two water molecules side by side, as shown in Figure 3. Further increase in distortion value results in a higher energy barrier, higher friction, and decreased pore volume. This reduces water transport significantly.

Figure 5 shows the comparison of water occupancy of the CNTs and BNNTs with different deformations. For the BNNT with the XY-distortion of 4, even though it has a larger pore volume compared to the CNT of the same distortion, very few molecules enter the pore. This shows that the entrance effects play a major role in water flow through nanotubes. On the visualization of the simulation, using the VMD package, we found that, for the XY-distortion of 4 and 6 (CNT and BNNT), molecules entered the nanotube, but were unable to travel further continuously. This is due to the large friction and energy barrier inside these nanotubes. For these cases, only a handful of molecules were observed to travel through the pore from one end to the other. The screw-distorted nanotubes showed a negligible effect on the water transport phenomena. They behaved almost similarly to the perfect nanotubes. This can be observed from the water occupancy and free energy of occupancy fluctuation plots.

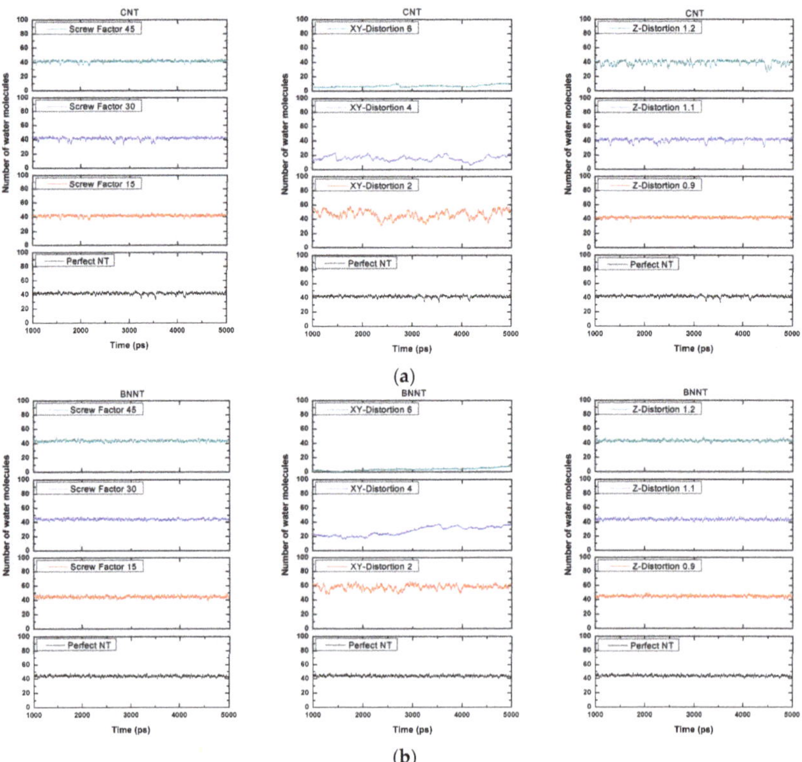

Figure 5. Water occupancy inside perfect and XY-distorted nanotubes versus time in picosecond: (**a**) carbon nanotube (CNT) and (**b**) boron nitride nanotube (BNNT).

3.2. Flux and Diffusion

Flux is created by the pressure that is applied to the water molecules inside the pool to push them through the nanotube. Water flux is calculated as the difference between the sum of the number of water molecules that move from the left side to the right side of the nanotube and the molecules which cross in the opposite direction [24]. Figure 6 shows the variation of flux based on different deformations.

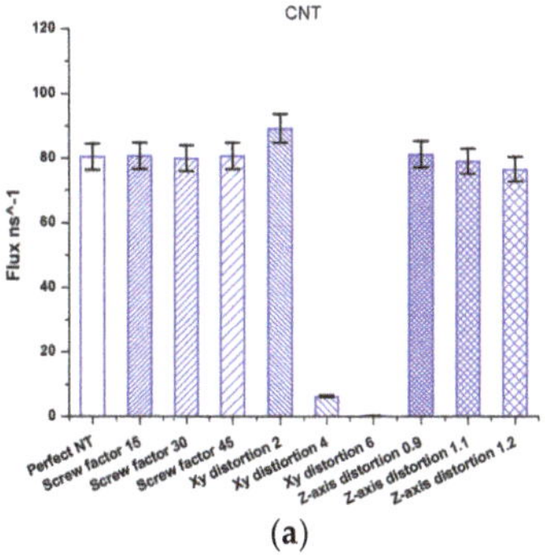

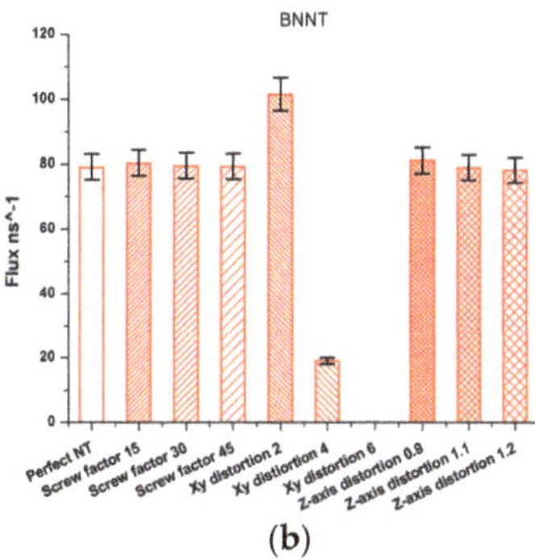

(**a**)　　　　　(**b**)

Figure 6. Variation of flux in perfect and different deformed nanotubes: (**a**) carbon nanotube (CNT) and (**b**) boron nitride nanotube (BNNT).

For both the carbon nanotube and boron nitride nanotube, the XY-distortion of 2 showed higher flux when compared to that of other nanotubes. XY-distortions of higher value showed much less flux when compared to the perfect nanotube. The number of molecules that cross the pore gradually decreases with an increase in the degree of deformation. The screw distorted nanotubes did not show any variations; they behave almost similarly to that of the perfect nanotube.

The diffusion coefficient of water through the nanotube is the measure of the mass of the water that diffuses through a unit surface of the nanotube in a unit time at a concentration gradient of unity. The axial diffusion coefficient of water molecules can be computed from the mean squared displacement (MSD). The MSD of the water molecule's center of mass can be calculated by using the relation used by Barati Farimani et al. [25]:

$$<|(r(t) - r(0))|^2 > = ADt^n$$

where 'r' denotes the center of the mass coordinate of the water molecule. The angle bracket used in the equation defines the average over all the water molecules; 't' denotes the time interval; 'D' denotes the diffusion coefficient; 'A' denotes the dimensional factor values of 2, 4, and 6 for 1-, 2-, and 3-dimensional diffusion, respectively; and 'n' defines the type of the diffusion mechanism. The value of 'n' can be 0.5, 1, and 2, depending on how the MSD varies with time. These values of 'n' represent Single-File diffusion, Fickian diffusion, and ballistic diffusion, respectively. MSD of all the molecules in that direction with A = 2 is used to compute the average axial diffusion coefficient in Z-direction. In this simulation, the single-file diffusion is observed in all cases other than the XY-distorted nanotubes. In the XY-distorted nanotube of distortion value 2, a change from single-file to near-Fickian diffusion is observed. Hence, there is a significant increase in diffusion and flux.

Figure 7 shows the variation of diffusion coefficient for nanotubes with different types of deformation. The XY-deformation of 2 shows a relatively higher diffusion coefficient due to the increase in the pore volume of the nanotube when compared to other XY-distortions of higher value. This is in good agreement with the Hilder et al. [26] and Corry et al. [27]. It can be seen that there are significant values for diffusion for XY-distortions of 4 and 6 when compared with that of the flux. This is due to the filling of water molecules near the entrance and exit of the nanotube, while the number of water molecules that completely travel from one end to the other is very low.

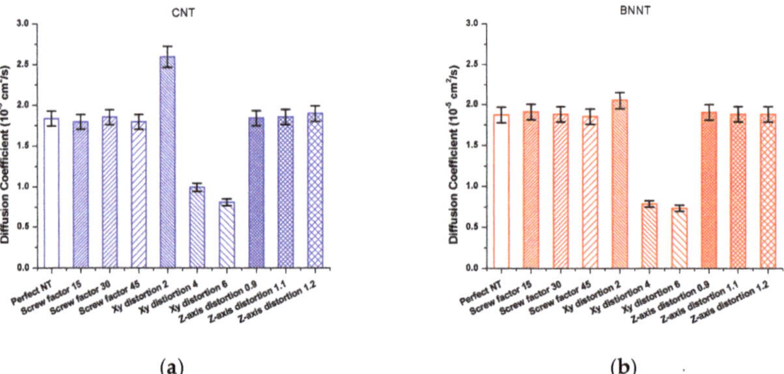

Figure 7. Pressure dependence of diffusion coefficient in perfect and different types of deformed nanotubes: (**a**) carbon nanotube (CNT) and (**b**) boron nitride nanotube (BNNT).

3.3. The Hydrogen Bonding

The average number of hydrogen bonds per water molecule inside the nanotube for the nanotubes with different deformations is given in the figure below for both BNNT and CNT.

We know that the number of hydrogen bonds that occurs inside the nanotube is a function of the length and the pore size. From Figure 8, we can observe that there is no significant increase in the number of hydrogen bonds formed inside the pore for the nanotubes, which has a twist, as they have the same length and pore size of that of the perfect nanotube. We observed that the nanotubes with the XY-distortion of 2 show a significantly higher value of hydrogen bonding per water molecule inside the nanotube when compared to others. This is due to the increase in the pore volume of the elliptical crosses section of the nanotubes, which facilitates the accommodation of water molecules side by side inside the pore. Usually, a single-file water transport is observed in nanotubes in (6,6) nanotubes, but when they are distorted in XY-direction, it has sufficient space to accommodate two water molecules side by side, as shown clearly in the Figure 3.

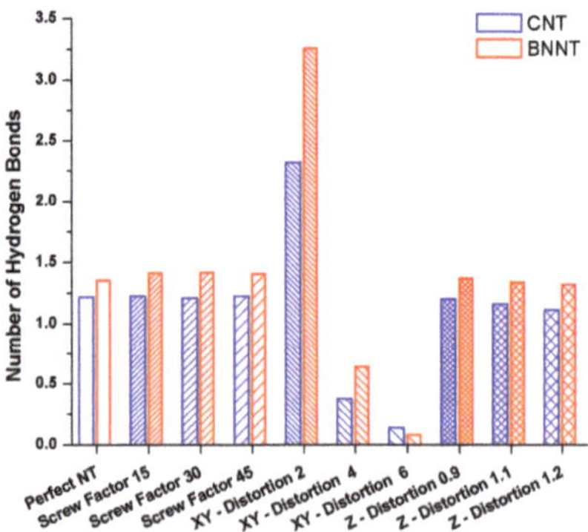

Figure 8. Average number of hydrogen bonds per water molecule formation inside the pore for perfect and different types of deformed nanotubes: carbon nanotubes (CNT) and boron nitride nanotubes (BNNT).

For both the CNT and BNNT, we can see a significant increase in the hydrogen bond, which decreases the mobility and formation of the more-bonded system. When compared between the diffusion and the number of hydrogen bond plots for the XY-distortion of 2, we can see that the diffusion in the CNT is higher with a lower hydrogen bond when compared with the BNNT. This shows that the formation of a more-bonded system decreases the water diffusion through the nanotubes. This is in good agreement with the results obtained by Mendonca et al. [28]. This facilitates the change of diffusion from single file transport to near-Fickian diffusion. The XY-distortion of 4 and 6 shows the accumulation of molecules near the entrance and exit of the pore. Due to the large energy barrier inside the pore and large interaction of the water molecules with the wall, the molecules can neither occupy the whole pore volume nor transport through them easily.

The Z-distortion of 1.2 of the nanotubes shows a noticeable decrease in the number of hydrogen bonds per water molecule when compared to other Z-distortions due to the stop-start diffusion of molecules near the entrance of the nanotube, as shown in the Figure 9. Even though this phenomenon is observed in both the CNT and BNNT, this phenomenon is predominant in the CNT when compared with that of the BNNT. To further understand this discontinuous flow phenomenon, the friction force, radial distribution function (RDF), and potential of mean force are studied for this particular case in detail for the CNT.

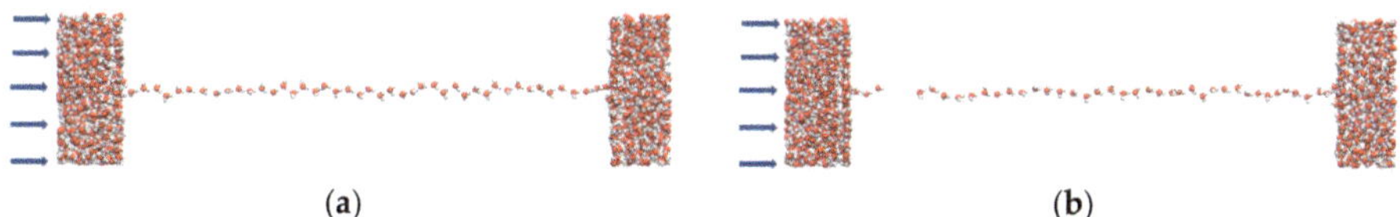

(a)   (b)

Figure 9. Visualization of water chain formed in-between two reservoirs: (**a**) perfect CNT and (**b**) CNT with Z-distortion value 1.2.

3.4. Friction Force and Trajectory

Discontinuous single-file diffusion is found in the CNT with the Z-distortion of 1.2, as shown in Figure 9. To investigate this phenomenon, the friction force is calculated. The friction force plays a significant role in water transport through the nanotubes. The friction force is large when there is a large interaction of water molecules with the nanotube walls.

Friction force is calculated by using the method suggested by Falk et al. [29]:

$$f(r) = dU(r) \, / \, dr = 24 \, (\varepsilon/\sigma) \, [(\sigma/r)^7 - (\sigma/r)^{13}]$$

where 'r' is the distance between the oxygen atom of the water molecule and the nanotube wall atoms. Figure 10a shows the plot of friction force along the nanotube length for the carbon nanotube with Z-distortion of 1.2 and the perfect nanotube. It can be seen that the Z-distortion of 1.2 shows a higher friction force when compared to the perfect nanotube. This implies that there is less interaction between the walls of the perfect nanotube with the water molecules when compared to that of the Z-distortion of 1.2.

The discontinuity of the single-file water transport arises near the entrance of the nanotube. It is either followed by breakage of the single-file transport into short chains of molecules or filled up again at a faster pace. The molecules that follow after the broken single-file chain move relatively faster when compared to that of other molecules inside the nanotube. This can be viewed in Figure 10b, which shows the plot of the trajectory of a single molecule inside the perfect CNT compared to that of the CNT with Z-distortion of 1.2.

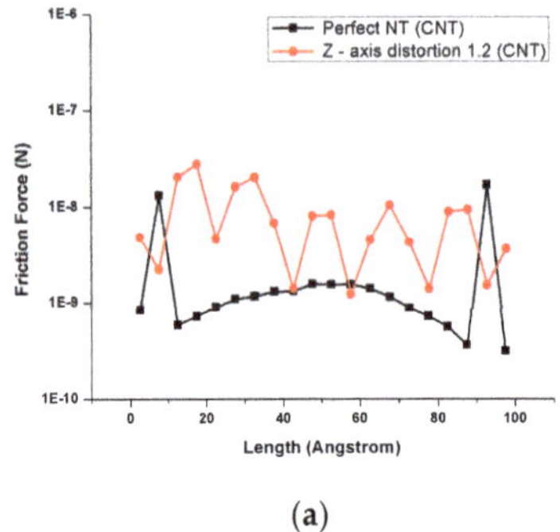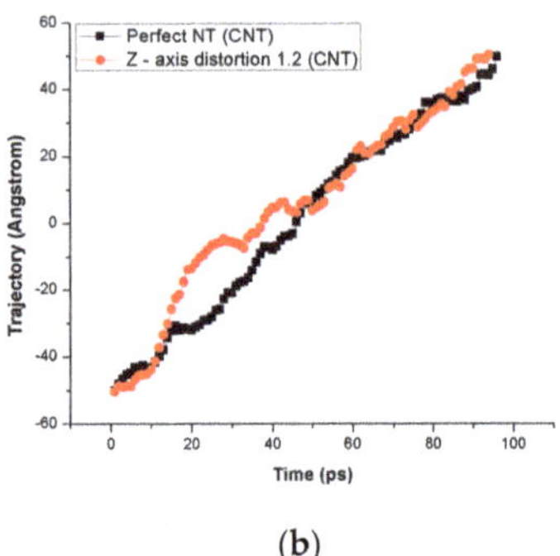

(a)   (b)

Figure 10. (**a**) Friction force inside the perfect nanotube and Z-distortion 1.2 CNT. (**b**) Comparison of water molecule trajectory inside perfect CNT and CNT with Z-distortion value 1.2.

3.5. Radial Distribution Function and Potential of Mean Force

The structural change of water molecules in the simulation cell can be described by the radial distribution function. Radial distribution function describes the atomic density variation as a function of distance from a particular atom [29]. Figure 11a illustrates the radial distribution function (RDF) of water molecules inside perfect and the Z-distortion 1.2 carbon nanotube.

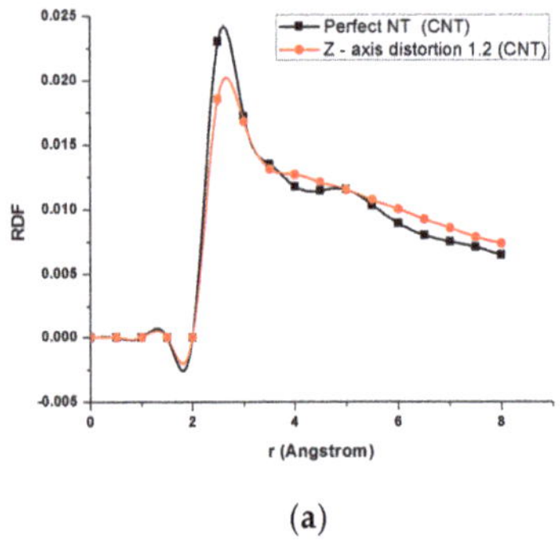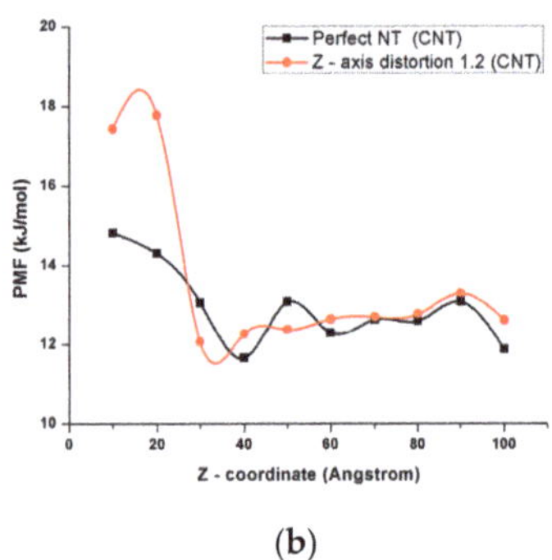

(a)   (b)

Figure 11. (**a**) Radial distribution function of water molecules within the perfect carbon nanotube and carbon nanotube with Z-distortion 1.2. (**b**) Potential of mean force within the perfect carbon nanotube and carbon nanotube with Z-distortion 1.2.

The Z-distortion of 1.2 shows the more favorable interplay between the water and walls of the nanotube. It shows that the density increases in the vicinity of the Z-distortion of 1.2 carbon nanotube walls when compared to that of the perfect carbon nanotube walls. Therefore, this strong interplay between the oxygen atom of the water molecules and the Z-distortion of 1.2 carbon nanotube walls reduces the transport rate of water inside this nanotube. When compared to that of the Z-distortion of the 1.2 carbon nanotube, perfect carbon nanotubes show less interplay between the oxygen atoms of the water molecule with the walls of the nanotube. Thus, it favors faster water transport when compared to the Z-distortion of 1.2.

Potential of mean force (PMF) calculations inside the nanotubes help to understand the amount of energy barrier that exists inside the nanotube [30]. Water molecules have to overcome this energy barrier to move through the nanotube. The potential of mean force is calculated as follows:

$$PMF = (-k_B T) \ln g(x)$$

where $g(x)$ is the radial distribution function (RDF), k_B the Boltzmann constant, and T the thermodynamic temperature [16]. Figure 11b shows that the potential of mean force inside the Z-distortion of 1.2 CNT is significantly larger than the perfect CNT, so the water molecules should have larger energy to conduct through the nanopore. This result further supports the reason for the discontinuous flow occurring inside the Z-distortion of the 1.2 carbon nanotube.

4. Conclusions

In conclusion, we have investigated the different types of deformations such, as screw distortion, XY-distortion, and Z-distortion in both carbon nanotubes and boron nitride nanotubes of 100 Å in length with chirality (6,6).

The effects of these deformations on both the nanotubes (BNNT and CNT) are quite similar. The formation of the more bonded system due to the increase in hydrogen bonds inside the nanotube decreases the diffusion of water through them. This is similar to the results reported by Mendonca et al. [28]. As Feng et al. [6] reported for the diffusion of helium gas through deformed nanotubes, the screw distortion did not have any significant impact on water transport, but the impact of the XY-distortion and Z-distortion were quite significant. The XY-deformation of higher value had a compelling negative effect on water transport, while the XY-distorted nanotubes of value 2 showed encouraging effects on water transport through the nanotubes. These results point to the importance of nanotube structure on water transport phenomena.

Further studies could be carried out for the XY-distortion of values less than 2 and for values slightly higher than 2 for optimizing the value which supports better water transport. These studies can also help to design better transmembrane channels and other relative nanodevices.

Author Contributions: Conceptualization, F.R.; methodology, F.R.; software, F.R.; validation, F.R., M.S. and D.K.; formal analysis, F.R.; investigation, D.K. and M.S.; resources, D.K.; data curation, F.R.; writing—original draft preparation, F.R.; writing—review and editing, M.S. and D.K.; visualization, F.R.; supervision, D.K. and M.S.; project administration, D.K. and M.S.; funding acquisition, D.K.

Funding: This research received no external funding.

Conflicts of Interest: The authors declare no conflict of interest.

References

1. Iijima, S. Helical microtubules of graphitic carbon. *Nature* **1991**, *354*, 56–58. [CrossRef]
2. Rubio, A.; Corkill, J.L.; Cohen, M.L. Theory of graphitic boron nitride nanotubes. *Phys. Rev. B* **1994**, *49*, 5081–5084. [CrossRef]
3. Chopra, N.G.; Luyken, R.J.; Cherrey, K.; Crespi, V.H.; Cohen, M.L.; Louie, S.G.; Zettl, A. Boron Nitride Nanotubes. *Science* **1995**, *269*, 966. [CrossRef]
4. Hummer, G.; Rasaiah, J.C.; Noworyta, J.P. Water conduction through the hydrophobic channel of a carbon nanotube. *Nature* **2001**, *414*, 188–190. [CrossRef]
5. He, J.-X.; Lu, H.-J.; Liu, Y.; Wu, F.-M.; Nie, X.-C.; Zhou, X.-Y.; Chen, Y.-Y. Asymmetry of the water flux induced by the deformation of a nanotube. *Chin. Phys. B* **2012**, *21*, 054703. [CrossRef]
6. Feng, J.; Chen, P.; Zheng, D.; Zhong, W. Transport diffusion in deformed carbon nanotubes. *Phys. A Stat. Mech. Appl.* **2018**, *493*, 155–161. [CrossRef]
7. Amiri, H.; Shepard, K.L.; Nuckolls, C.; Hernández Sánchez, R. Single-Walled Carbon Nanotubes: Mimics of Biological Ion Channels. *Nano Lett.* **2017**, *17*, 1204–1211. [CrossRef] [PubMed]
8. Liu, B.; Li, X.; Li, B.; Xu, B.; Zhao, Y. Carbon Nanotube Based Artificial Water Channel Protein: Membrane Perturbation and Water Transportation. *Nano Lett.* **2009**, *9*, 1386–1394. [CrossRef] [PubMed]
9. Gong, X.; Li, J.; Lu, H.; Wan, R.; Li, J.; Hu, J.; Fang, H. A charge-driven molecular water pump. *Nat. Nanotechnol.* **2007**, *2*, 709–712. [CrossRef] [PubMed]

10. Wan, R.; Li, J.; Lu, H.; Fang, H. Controllable Water Channel Gating of Nanometer Dimensions. *J. Am. Chem. Soc.* **2005**, *127*, 7166–7170. [CrossRef] [PubMed]
11. Li, J.; Gong, X.; Lu, H.; Li, D.; Fang, H.; Zhou, R. Electrostatic gating of a nanometer water channel. *Proc. Natl. Acad. Sci. USA* **2007**, *104*, 3687–3692. [CrossRef] [PubMed]
12. Kalra, A.; Garde, S.; Hummer, G. Osmotic water transport through carbon nanotube membranes. *Proc. Natl. Acad. Sci. USA* **2003**, *100*, 10175–10180. [CrossRef] [PubMed]
13. Zhu, F.; Schulten, K. Water and Proton Conduction through Carbon Nanotubes as Models for Biological Channels. *Biophys. J.* **2003**, *85*, 236–244. [CrossRef]
14. Todorov, I.T.; Smith, W.; Trachenko, K.; Dove, M.T. DL_POLY_3: New dimensions in molecular dynamics simulations via massive parallelism. *J. Mater. Chem.* **2006**, *16*, 1911–1918. [CrossRef]
15. Bush, I.J.; Todorov, I.T.; Smith, W. A DAFT DL_POLY11URL distributed memory adaptation of the Smoothed Particle Mesh Ewald method. *Comput. Phys. Commun.* **2006**, *175*, 323–329. [CrossRef]
16. Humphrey, W.; Dalke, A.; Schulten, K. VMD: Visual molecular dynamics. *J. Mol. Graph.* **1996**, *14*, 33–38. [CrossRef]
17. Zhu, F.; Tajkhorshid, E.; Schulten, K. Theory and simulation of water permeation in aquaporin-1. *Biophys. J.* **2004**, *86*, 50–57. [CrossRef]
18. Werder, T.; Walther, J.H.; Jaffe, R.L.; Halicioglu, T.; Koumoutsakos, P. On the Water–Carbon Interaction for Use in Molecular Dynamics Simulations of Graphite and Carbon Nanotubes. *J. Phys. Chem. B* **2003**, *107*, 1345–1352. [CrossRef]
19. Raghunathan, A.V.; Park, J.H.; Aluru, N.R. Interatomic potential-based semiclassical theory for Lennard-Jones fluids. *J. Chem. Phys.* **2007**, *127*, 174701. [CrossRef]
20. Tang, D.; Li, L.; Shahbabaei, M.; Yoo, Y.-E.; Kim, D. Molecular Dynamics Simulation of the Effect of Angle Variation on Water Permeability through Hourglass-Shaped Nanopores. *Materials* **2015**, *8*, 7257–7268. [CrossRef]
21. Won, C.Y.; Aluru, N.R. Water Permeation through a Subnanometer Boron Nitride Nanotube. *J. Am. Chem. Soc.* **2007**, *129*, 2748–2749. [CrossRef] [PubMed]
22. Ryckaert, J.-P.; Ciccotti, G.; Berendsen, H.J.C. Numerical integration of the cartesian equations of motion of a system with constraints: Molecular dynamics of n-alkanes. *J. Comput. Phys.* **1977**, *23*, 327–341. [CrossRef]
23. Shahbabaei, M.; Kim, D. Molecular Dynamics Simulation of Water Transport Mechanisms through Nanoporous Boron Nitride and Graphene Multilayers. *J. Phys. Chem. B* **2017**, *121*, 4137–4144. [CrossRef] [PubMed]
24. Shahbabaei, M.; Kim, D. Effect of hourglass-shaped nanopore length on osmotic water transport. *Chem. Phys.* **2016**, *477*, 24–31. [CrossRef]
25. Barati Farimani, A.; Aluru, N.R. Spatial Diffusion of Water in Carbon Nanotubes: From Fickian to Ballistic Motion. *J. Phys. Chem. B* **2011**, *115*, 12145–12149. [CrossRef] [PubMed]
26. Hilder, T.A.; Gordon, D.; Chung, S.-H. Salt Rejection and Water Transport Through Boron Nitride Nanotubes. *Small* **2009**, *5*, 2183–2190. [CrossRef]
27. Corry, B. Designing Carbon Nanotube Membranes for Efficient Water Desalination. *J. Phys. Chem. B* **2008**, *112*, 1427–1434. [CrossRef]
28. Mendonça, B.H.S.; de Freitas, D.N.; Köhler, M.H.; Batista, R.J.C.; Barbosa, M.C.; de Oliveira, A.B. Diffusion behaviour of water confined in deformed carbon nanotubes. *Phys. A Stat. Mech. Appl.* **2019**, *517*, 491–498. [CrossRef]
29. Falk, K.; Sedlmeier, F.; Joly, L.; Netz, R.R.; Bocquet, L. Ultralow Liquid/Solid Friction in Carbon Nanotubes: Comprehensive Theory for Alcohols, Alkanes, OMCTS, and Water. *Langmuir* **2012**, *28*, 14261–14272. [CrossRef]
30. Won, C.Y.; Joseph, S.; Aluru, N.R. Effect of quantum partial charges on the structure and dynamics of water in single-walled carbon nanotubes. *J. Chem. Phys.* **2006**, *125*, 114701. [CrossRef]

Article

Experimental Investigation of Copper Mesh Substrate with Selective Wettability to Separate Oil/Water Mixture

Jia Yuan [1,2], Chenyi Cui [1], Baojin Qi [1,2,*], Jinjia Wei [1,3] and Mumtaz A. Qaisrani [3]

[1] School of Chemical Engineering and Technology, Xi'an Jiaotong University, Xi'an 710049, China; yj2551241405@stu.xjtu.edu.cn (J.Y.); ccy8600@stu.xjtu.edu.cn (C.C.); jjwei@xjtu.edu.cn (J.W.)
[2] Xi'an Jiaotong University Suzhou Academy, Suzhou 215123, China
[3] State Key Laboratory of Multiphase Flow in Power Engineering, Xi'an Jiaotong University, Xi'an 710049, China; mumtazqaisrani@yahoo.com
*   Correspondence: bjqi@mail.xjtu.edu.cn

Received: 20 October 2019; Accepted: 21 November 2019; Published: 29 November 2019

Abstract: To solve the problem of low efficiency and poor adaptability during complex oil/water mixtures separation, two types of membranes with superhydrophilicity/underwater-superoleophobicity were successfully fabricated by oxidative reaction and in situ displacement reaction methods. A nanoneedle $Cu(OH)_2$ structure was generated on the copper mesh substrate by oxidative reaction and feathery micro/nanoscale composite, while Ag structure was constructed at the surface of copper mesh substrate through in-situ replacement, then, membranes with superhydrophilic/ underwater-superoleophobic properties were separated. The influence of microstructure, wettability of the surface of prepared membranes and the liquid constituents in the separation experiment were studied and the liquid flux and permeation pressure at the membrane were later experimentally investigated. The experimental results show that separation efficiency of both membranes for separating different oil/water mixtures was above 99.8%. However, the separation efficiency of the Ag-CS (Ag on the copper substrate) membrane was obviously higher than that of the $Cu(OH)_2$-CS ($Cu(OH)_2$ on the copper substrate) membrane after 10 instances of separation because of the micro/nanocomposite structures. By comparison, it was found that the Ag-CS membrane showed a relatively higher permeation pressure but lower liquid flux as compared to $Cu(OH)_2$-CS membrane, due to the influence of microscale structure and the wettability of the surface combined. In addition, the outcome for separating the multicomponent oil/water mixture illustrate that the result of TOC (the Total Organic Carbon) test for the $Cu(OH)_2$-CS membrane and Ag-CS membrane were 31.2% and 17.7%, respectively, higher than the average of the two oils probably because some oil droplets created due to mutual dissolution passed through the membranes. However, these two fabricated membranes still retained higher separation efficiencies and good adaptability after 10 instances of separation. It was concluded that based on the good performances of the prepared membranes, especially the modified membrane, they have a vast application prospect and can be widely used.

Keywords: oil/water separation; superhydrophilic/underwater-superoleophobic membranes; opposite properties; superhydrophobicity/superoleophilicity; selective wettability; micro/nanoscale composite structure

---

1. Introduction

During recent years, large discharge quantities of oily wastewater has attracted the attention of the public, and it also has drawn researchers' interests in the field of fabricating novel materials having higher oil/water separation efficiency. There are two primary ways to generate oily wastewater: the first is by releasing the oil-contaminated industrial wastewater; these industries often constitute of petrochemical industries, printing industries, metallurgical-production industries, food-processing industries, and so on [1–6]. Such wastewater has become the most common contaminant all over the world and it has seriously threatened our habitat. The other reason for oily wastewater generation is the large numbers of oil spill incidents and the recent estimates show that nearly two million tons of oil spills into the ocean annually [5–7]. So, a significant quantity of waste oil being disposed into the oceans has not only caused substantial energy loss, but also seriously threatened aquatic life because the spilled oil decreases the oxygen present in the water [8–13]. In addition, the fouling of surfaces is a significant problem that affects not only people' daily lives, but also does harm to industrial production systems, and it could be mitigated via a method by combining particular designed interfaces and further chemical treatments. This is a good technique with broad application prospects.

If the membranes with low surface adhesion to contaminants on the surface could be prepared, it will greatly benefit the lifetime and separation efficiency of the membranes in practical applications [14]. At present, several traditional methods including ultrasonic separation, skimming, centrifugation, combustion, etc., are being widely applied for the oily wastewater separation [9,15–18]. Nevertheless, the application range of these techniques is often hampered by some limitations such as the production of secondary pollutants, low efficiency, and the complex operation [3,15,19–21]. Considering the issues above, there's an urgent requirement for novel technologies having characteristics, higher separation efficiency, low operational cost and simple operation for separation of oil/water mixture.

Inspired by the naturally occurring phenomenon of superhydrophobicity at the surface of lotus leaf and goose feathers, scholars have researched and fabricated new membrane-based materials with high separation efficiency of oil/water mixtures. These membrane materials with unique wetting properties have gradually become a hot research topic in the past decade [6,22–24]. Wettability exists as an inherited characteristic of solid surfaces, and it affects the wetting phenomenon as the droplets touch the surface of the solid [25–28]. According to the different wetting phenomena when oil droplets and water droplets touch the solid surface, it can be summarized into four wetting properties, namely: oleophilicity, oleohobicity, hydrophilicity and hydrophobicity. Researchers found that the hierarchical micro/nanostructures could increase the roughness of the solid surface and then enhanced the four-fundamental wetting properties into superoleophilicity, superoleohobicity, superhydrophilicity and superhydrophobicity [29–34] and some organics with lower surface energy could decrease the wettability of liquid and solid surfaces. By employing physical refining as well as chemical treatment approaches such as corrosion, alteration of some substances with lower energy, electric deposition etc., to formulate the micro and nanocomposite structure membrane with relatively lower energy surfaces [20,35]. On the basis of above analyses the membrane-based separation material driven solely by gravity can be divided into two categories: "water-removing" and "oil-removing". The first type material called "water-removing" membrane means the membrane shows "water-loving" and "oil-hating" characteristics, and water easily passes across the membrane; however, oil is blocked over the membrane simultaneously. On the contrary, the membrane of the "oil-removing" material shows "oil-loving" and simultaneously "water-hating" characteristics, and water stops above the membrane while oil can permeate the membrane smoothly [36–39].

Researchers first prepared the superhydrophobic/superoleophilic surface, which could selectively let the oil pass through but stopped the water and the material successfully separated the heavier oil/water mixture [9,24]. However, there are two main problems associated with the practical application of the superhydrophobic/superoleophilic membranes: on the one hand, this material is suitable to separate the heavier oils/water, like 1,2-dichloroethane, whose density is heavier than the density of water, but oils having a density lower than water; it becomes complex to practically separate the mixture because the oil cannot touch the membrane even though it could easily pass across the membrane, and is occluded by the lower water layer. In such a situation, placing the separation device obliquely is necessary to separate the mixture smoothly [38–40]. While, on the other hand, as a result of higher viscosity, the oil phase quickly chokes the membrane during the separation. Thus, the separation rate can be much lower, and that limits the practical application in the field of oil/water separation [41–43]. Considering the problems above, scholars have investigated the superhydrophilicity/superoleophobicity phenomenon compared with the superhydrophobicity/superoleophilicity of the oil-removing-type material.

After continuous exploration and attempts by many scholars, this material was successfully fabricated [44]. The "water-removing" membrane could ideally avoid the two problems mentioned above to separate oil with a lower density than the water, as water smoothly flows across the superhydrophilic membrane and oil stays above the superoleophobic membrane. Even if it perfectly avoids these two problems, there still remain a few disadvantages of the superhydrophilic/superoleohobic materials. Many "water-removing" materials need to have characteristics with lower surface energy to further modify its solid surface, but actually those materials with lower surface energy are usually fluorine-containing organic matter [23,31,45,46]. In that case, the formation process is usually complex and costly, but also the chemical stability of the adapted surface coating is not very effective. Furthermore, the literature shows that the current research hitherto primarily focused on separating the single oil and water mixture. However, it should be noted that the composition of wastewater containing oil varies and was often complex due to presence of different oil components; this certainly yields a high demand for membrane adaptability to the practical applications of oil/water separation [8,15,36].

The current study reports two facile and one-step chemical reaction methods on the copper mesh to fabricate the "water-removing" membrane with superhydrophilicity and underwater-superoleophobicity. The prepared membranes possess a nanoneedle $Cu(OH)_2$ structure and a micro/nanocomposite Ag structure. With these two easily operable methods, we successfully fabricated the water-removing membranes and experimentally studied the separation efficiency, permeation pressure and fluid flux. In addition, to authenticate the more widespread adaptability of the membranes, the mixture with a single component of oil and water were successfully separated, and the multi-constituents of the oil/water mixture were experimentally investigated.

2. Experimental Platform

2.1. Materials

Four-hundred copper mesh substrates were used as obtained. Deionized water was self-prepared. Other reagents including $AgNO_3$, HCl, Oil Red O (Sudan) etc. from Sinopharm Chemical Reagents Group were of analytical grade and were used without any further purification. Diesel and gasoline were purchased from Sinopec, and sunflower oil was purchased from the supermarket.

2.2. Sample Preparation

The copper meshes were sized into 6×6 cm^2 fragments for the following use. To make clean membranes, acetone along with ethanol and deionized water in beakers was used to serially soak all the meshes. Beakers were then left in an ultrasonic cleaner for nearly 10 min. In the end, the meshes were dried to remove any remaining contaminants at the surface of the substrate. To eliminate contaminants at the surface of the substrate, the meshes were later dried.

To prepare the $Cu(OH)_2$-CS membrane, firstly placing a cleaned copper mesh prepared as above into the 1 M NaOH and 1 M $K_2S_2O_8$ mixed solution horizontally for almost an hour, the prepared membrane was then rinsed with deionized water and dried at room temperature. Then, the $Cu(OH)_2$-CS membrane was obtained. To fabricate the Ag on the copper substrate (Ag-CS) membrane, 100 mL 0.1 M $AgNO_3$ solution was first prepared, and a piece of the cleaned copper mesh into the solution was added for 20 s at room temperature, then using the deionized water to washed the membrane for 3 min to clear out the residue and it was then dried in the blast drying oven at 60 °C.

2.3. Instrumentation and Characterization

The images of the SEM (the Scanning Electron Microscope) could be gained via a field-emission scanning electron microscope MAIA3 LMH and the images of XRD (the X-ray Diffraction) could be attained via an X-ray diffraction-6100. The images of XPS were gained by using a Thermo Fisher ESCALAB Xi^+ Spectrometer with the help of an Al $K\alpha$ X-ray source. Contact angles of membranes toward oil or water were measured by using the DSA100 machine at room temperature. The volume of the tested liquid (water or oil droplets) was 4 µL. To measure the water and oil contact angles, the droplets were positioned directly over the surface of the membrane and in the air, respectively. During the underwater oil contact angle measurement, 1,2-dichloroethane was chosen as tested oil due to its high density as compared to water and the membranes were secured in a water-filled transparent quartz vessel during the measurement. We used the average value of three measurements carried out at different locations on the one membrane sample to characterize the contact angle of the membrane. The oil content in the water after the separation was measured by the Total Organic Carbon Analyzer (vario TOC, Xi'an West Economics Import & Export Corporation, Xi'an, China).

3. Characterizations and the Separation Results of the Membranes

3.1. Microstructure on the Surface of the Membranes

The SEM images of the $Cu(OH)_2$-CS membrane as well as the Ag-CS membrane are shown in Figure 1. It can be inferred from Figure 1a,b that the pore diameter of the mesh is about 40 µm and the wire of copper mesh as substrate possessed a glossy surface—the oil smoothly flowed through the hole of such diameter. Figure 1c,d show the prepared membranes having rough surface structures. After reacting with 1 M $K_2S_2O_8$ and 1 M NaOH mixed solution, the surface of the $Cu(OH)_2$-CS membrane has an interval nanoneedle structure (Figure 1c,e)" Figure 1d,f show the feathery composite structure with the micronanoscales over the external surface of the Ag-CS membrane, which was made by immersing the original copper mesh in the 0.1 M $AgNO_3$ solution for about 20 s. We have tried several times to fabricate the Ag-CS membrane with an Ag structure generated on the surface of the membrane, but the results were similar with that showed in Figure 1d. The reason for the nonuniform-generated coating is the length of the reaction time.

In comparison, we can obviously see the difference between the nanoneedle structure generated on the surface of $Cu(OH)_2$-CS membrane and the feathery structure with micronanoscales on the Ag-CS membrane. The wettability of the solid surface is determined by both physical and chemical factors, such as the roughness of the surface and the hydrophobicity of some of the groups existing on the solid surface, respectively. The hierarchical feathery micro/nanostructure provides considerable roughness on the membrane surface compared with the nanoneedle structure, which may be more advantageous for the selective wettability of the membranes because it enhances the hydrophilicity in air and underwater-oleophobicity in water as per the Wenzel model [47]. For the mechanical strength of the generated coatings on the surface of membranes, Figure 1g,h demonstrate that the coatings' thickness on the membranes' surfaces was nearly 10 um and ensured the reusability of the membranes.

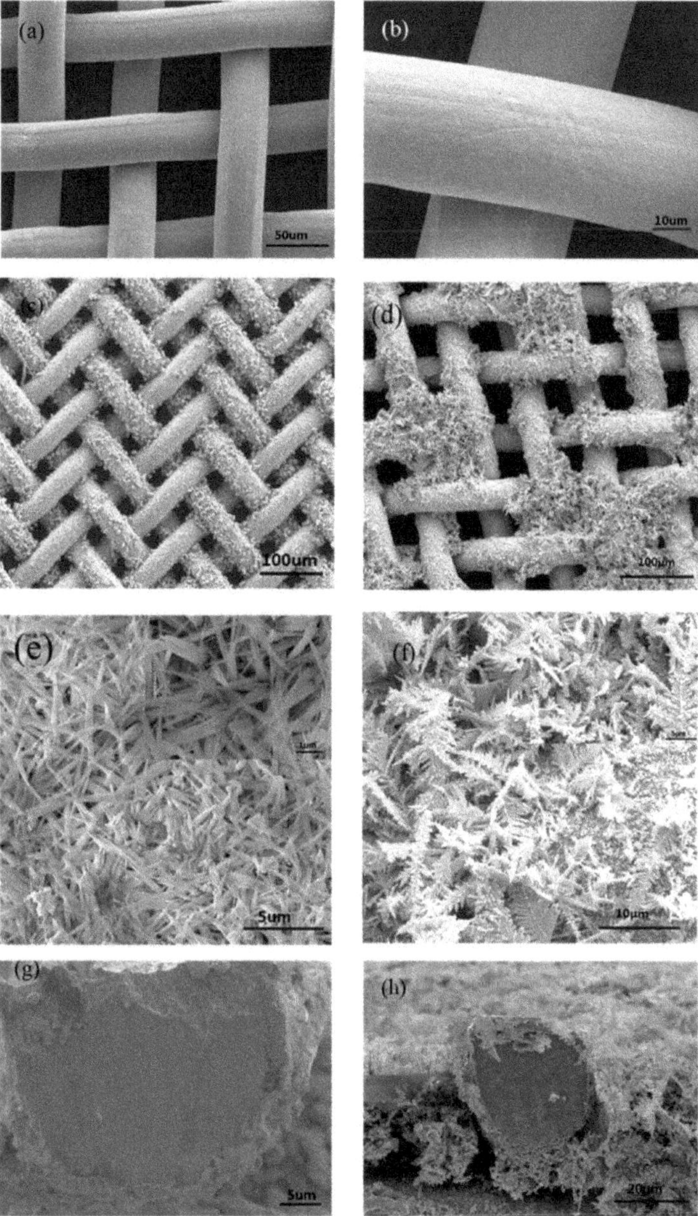

Figure 1. SEM images of the copper mesh substrate and the two prepared membranes: (**a**) SEM (the Scanning Electron Microscope) image of the original copper mesh; (**b**) a high-magnification image the copper mesh as substrate before chemical reaction; (**c**) the SEM image with low magnification of Cu(OH)$_2$-CS membrane; (**d**) the SEM image with low magnification of Ag-CS membrane; (**e**) the Cu(OH)$_2$-CS membrane displaying the generated nanoneedle structure; the insert is also the interval nanoneedle structure with high magnification; (**f**) the Ag-CS membrane displaying the generated micro/nano feathery structure; the insert is also the micro/nano feathery structure with high magnification; (**g**) the SEM image of the coating generated on a wire of the Cu(OH)$_2$-CS membrane; (**h**) the SEM image of the coating generated on a wire of the Ag-CS membrane.

3.2. The Composition of the Prepared Membranes

The crystal structures were validated by XRD of both the prepared membranes as depicted in Figure 2a,b. The letter "a" marked the Cu diffraction peaks in Figure 2a,b according to the number 04-0836 in the JCPDS card. In Figure 2a the peaks marked with the letter "b" represent $Cu(OH)_2$ according to the number 13-0420 in the JCPDS card and the letter "c" marked the Ag peaks in Figure 2b according to the number 89-3722 in the JCPDS card. Figure 2a clearly shows that the chemical composition of the generated structure on $Cu(OH)_2$-CS membrane is primarily $Cu(OH)_2$, while it can also be seen in Figure 2b that the letter "c" is marked Ag-generating on the Ag-CS membrane. The creation of the nanoscale $Cu(OH)_2$ structure on the first membrane and the micro/nanoscale feathery Ag structure on the other membrane are due to the following chemical reaction:

$$Cu + K_2S_2O_8 + 2NaOH \rightarrow Cu(OH)_2 + Na_2SO_4 + K_2SO_4 \tag{1}$$

$$Cu + AgNO_3 \rightarrow Cu(NO_3)_2 + Ag \tag{2}$$

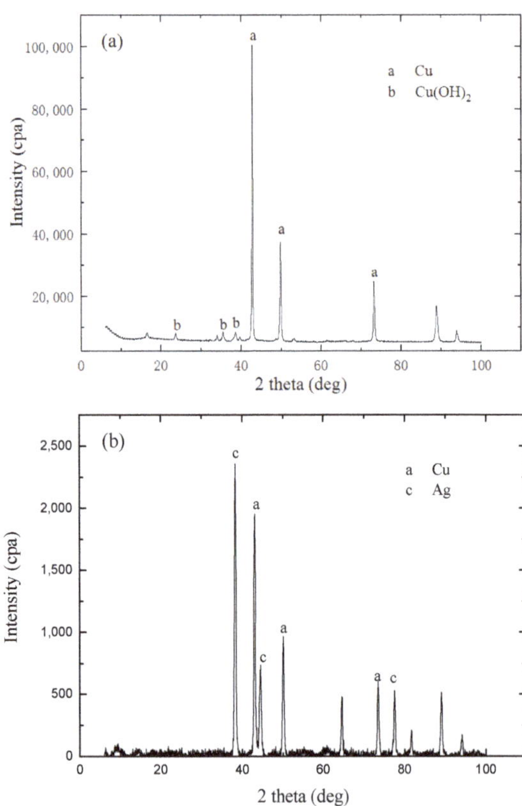

Figure 2. XRD (the X-ray Diffraction) images of the prepared membranes: (**a**) XRD image of the $Cu(OH)_2$-CS membrane; (**b**) XRD image of the Ag-CS membrane.

In Reaction (1), an oxidation reaction occurred, and the $K_2S_2O_8$ oxidizes the Cu substrate to $Cu(OH)_2$ raised at the external surface of the copper mesh. That is the chemical development process for fabrication of the $Cu(OH)_2$-CS membrane. In Reaction (2), the C9u substrate was oxidized to Cu^{2+} while Ag^+ in the 0.1 M $AgNO_3$ solution was reduced to Ag and then raised over the substrate surface.

The product $Cu(OH)_2$ and Ag appeared on the two membranes due to the above mentioned two chemical reactions is consistent with the outcomes of the XRD testing.

To further verify the structure's chemical composition generated over the membranes' surfaces, the results of XPS test in Figure 3a,b are also shown. From Figure 3a, it can be found that Cu^{2+} lies on the $Cu(OH)_2$-CS membrane, indicating that a portion of Cu substrate was oxidized. XRD test results as shown in Figure 2a, along with these findings, indicate that the generated structure on the surface of $Cu(OH)_2$-CS membrane is $Cu(OH)_2$. Figure 3b clearly displays that Ag exists on the surface of the Ag-CS membrane and this conclusion is in line with the outcome of XRD test. In this case, considering with the test outcomes of XRD shown in Figure 2b, it can be determined that the generated feathery micro/nanocomposite structure on the Ag-CS membrane surface is Ag.

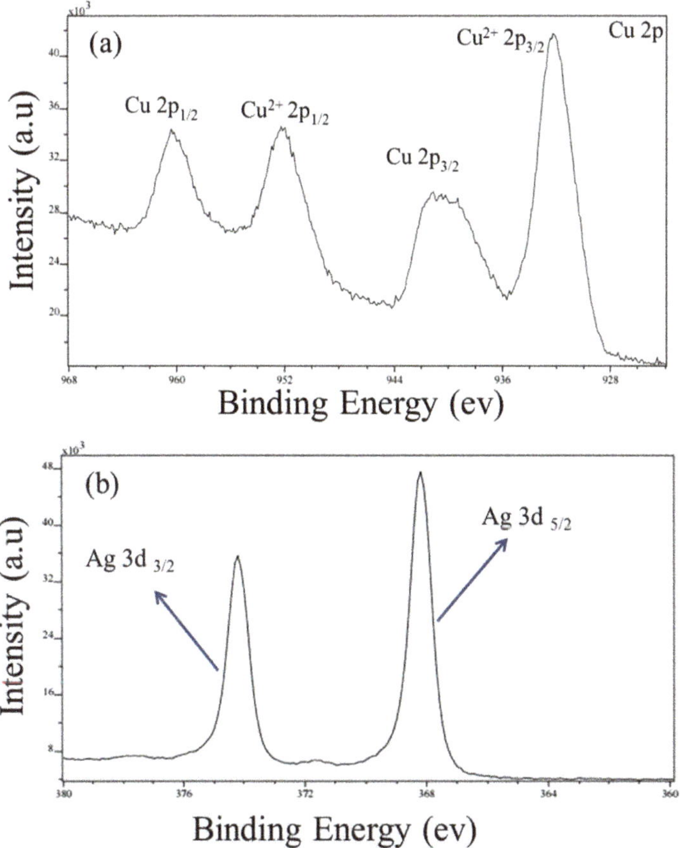

Figure 3. XPS (X-ray Photoelectron Spectrometer) images of the fabricated membranes: (**a**) XPS image of the $Cu(OH)_2$-CS membrane; (**b**) XPS image of the Ag-CS membrane.

3.3. Wettability of the Prepared Membranes

The oil and water contact angle were used to characterize the wettability of the membrane surface. When measuring the water and oil contact angles in the air, water droplet and oil droplet were released directly onto the membranes, when using the underwater oil contact angle to characterize the underwater-superoleophobicity of membranes, which is fixed in transparent quartz filled with water. Figure 4(a1,a2) show the diesel droplet and water droplet contact with the unreacted copper mesh substrate in air; the oil contact angle (OCA) and the water contact angle (WCA) were nearly 0° and 78.1°, respectively, and the unreacted copper mesh showed superoleopholicity and hydrophilicity. Figure 4b,c clearly displayed the $Cu(OH)_2$-CS membrane and Ag-CS membrane were superlyophilic in air as they come in contact with water and diesel, also, the contact angles were close to 0°. From Figure 4(a2), Figure 1e,f, we can imply that these two membranes both have a rather high roughness. The Wenzel model [47] also suggests that the rather high roughness of surfaces turn the hydrophilicity in air to superhydrophilicity of the membranes. For the same reason, water comes in contact with the membranes and the oil contact angle was nearly 0°. Moreover, when touching other oil types like gasoline and hexane etc., these two types of prepared membranes also displayed superoleophilicity performance in the air, which means such oil types can smoothly spread onto the membranes.

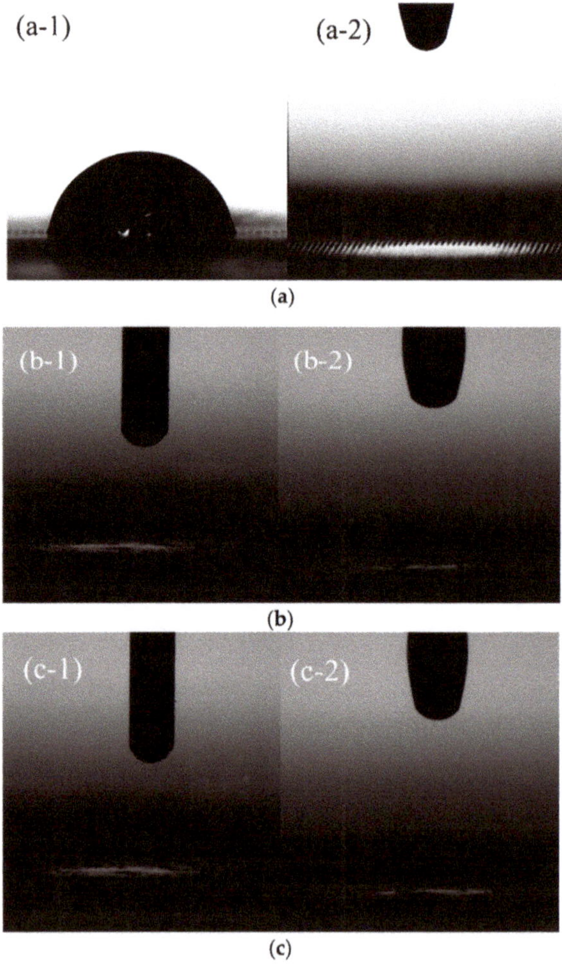

Figure 4. The contact angles of water and oil droplets when touching the original mesh and prepared membranes in the air. (**a-1**) The contact angle of water droplet on the surface of the unreacted copper mesh substrate in the air; (**a-2**) the contact angle of oil droplet on the surface of the unreacted copper mesh in the air; (**b-1**) the contact angle of a water droplet on the surface of the $Cu(OH)_2$-CS membrane in the air; (**b-2**) the contact angle of an oil droplet on the surface of the $Cu(OH)_2$-CS membrane in the air; (**c-1**) the contact angle of a water droplet on the surface of the Ag-CS membrane in the air; (**c-2**) the contact angle of an oil droplet on the surface of the Ag-CS membrane in the air.

Since the sym-dichloroethane is heavier than water, sym-dichloroethane was chosen as the tested oil to measure the underwater oil contact angle (UWOCA). Figure 5a,b show the superoleophobic performance of tested oil over the Cu(OH)$_2$-CS and Ag-CS membranes underwater and the underwater oil contact angles of these two prepared membranes i.e., 160.3° and 156.7°, respectively. The large contact angle values indicate that both membranes exhibited strong rejection toward oil droplets underwater and the membranes exhibited underwater-superoleophobicity. This may have occurred due to the fact that the superhydrophilic membranes surfaces were employed by water and its molecules were trapped inside of the generated, rather rough, structures on the surface and made the whole three-phase system achieve a state with relatively lower energy. The gas–liquid–solid (gas–oil–surface) interface on the membrane surface was changed into a liquid–liquid–solid (water–oil–surface) interface. As the membranes touched the oils, the oil molecules were unable to travel into the pore channels of membranes because the membranes blocked the oil molecules from swapping with water molecules and disrupt the relatively stable state with lower energy; in this case, the membranes naturally behaved with superhydrophilicity and underwater-superoleophobicity.

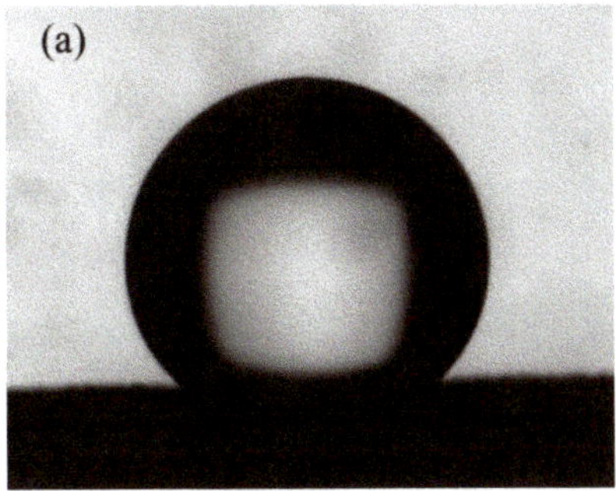

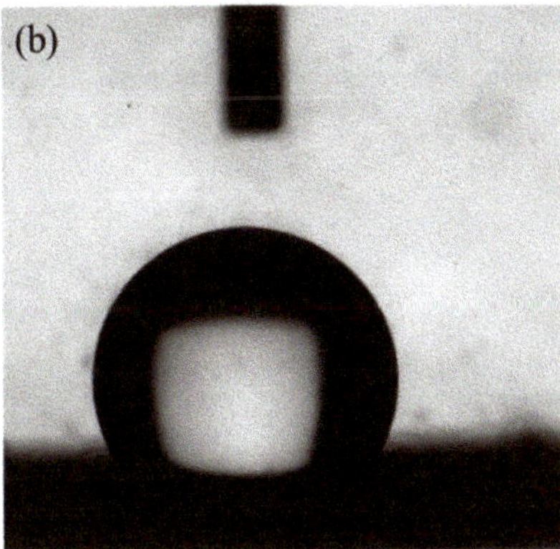

Figure 5. Underwater oil contact angles of the prepared membranes: (**a**) the superoleophobic performance of a sym-dichloroethane droplet when in contact with the Cu(OH)$_2$-CS membrane underwater; (**b**) the superoleophobic performance of a sym-dichloroethane droplet when in contact with the Ag-CS membrane underwater.

3.4. Single Component Oil/Water Mixture Separation

Diesel was chosen as the tested oil to carry out the single component oil/water mixture separation experiment. Figure 6 displays the photograph of the separation process, including before separation, in separation and after separation. From Figure 6, it can be seen that the prepared membranes were clamped in-between the two glass tubes, with each tube have two hooks and the tubes were fastened using four rubber bands to ensure the membranes were held firmly. Because the membranes displayed superhydrophilicity simultaneously with underwater-superoleophobicity and the tested oil diesel was lighter than water in the separation, the separation device was placed vertically. After being wetted beforehand, the water smoothly passed through the membrane while the diesel was repelled and stayed above the membrane. When the mixture consisting of water and diesel (dyed with Oil Red O) was poured, the water flowed across the membrane swiftly due to gravity, while diesel was stopped and remained above the membrane (please refer to the video in the Supplementary Information).

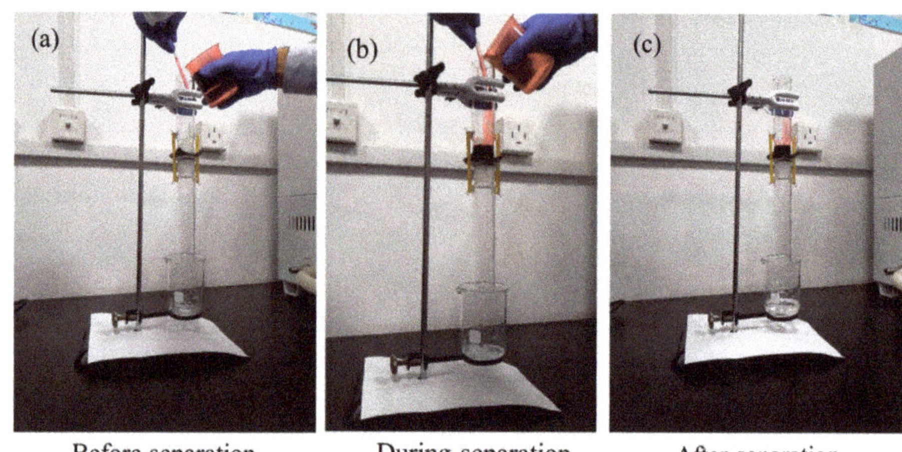

Figure 6. The image of separation process: (**a**) the image of the process before separation; (**b**) the image of the process during separation; (**c**) the image of the process after separation.

The separation of oil/water mixtures consisting of hexadecane/water or hexane/water or gasoline/water etc. were also carried out successfully. There was no visible oil (dyed with Oil Red O) in the water after the separation was carried out, while the constituents of the oil were tested by the TOC (the Total Organic Carbon). The efficiency of the separation was evaluated using following equation:

$$R(\%) = \left(1 - \frac{m_0}{m_c}\right) * 100\% \tag{3}$$

In the equation, m_0 represents the result of the TOC value, and m_c denotes the oil content of the oil/water mixture before separation. Figure 7a,b clearly display the results of TOC values and the calculated separation efficiency of the prepared membranes, respectively. Figure 7a,b show that the separation efficiency of the prepared membranes decreased with increasing reuse time, nevertheless, two membranes still performed well and kept high separation efficiency beyond 99.7% after 10 instances of separation. In the experiment, the used membranes were soaked in ethanol for about 5 min. Compared with the $Cu(OH)_2$-CS membrane, the Ag-CS membrane showed higher efficiency as well as better reusability in separating the oil/water mixture.

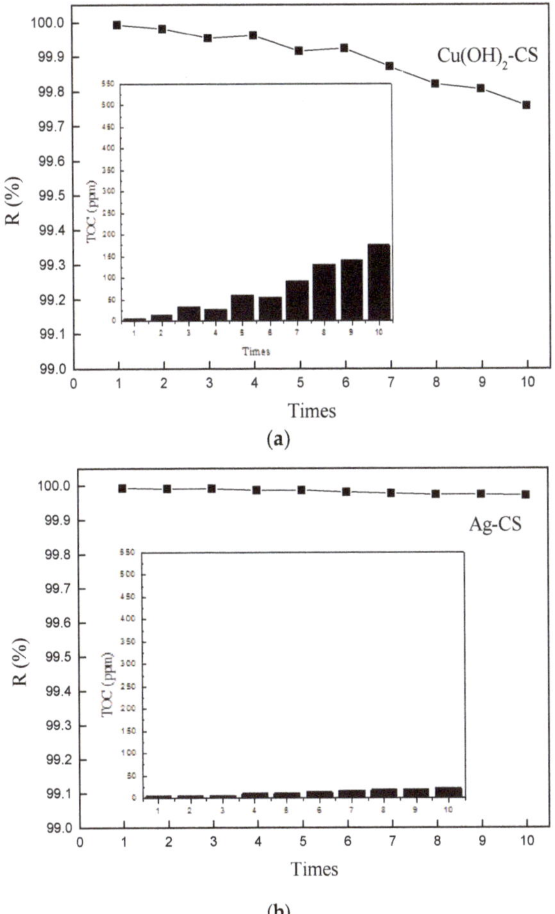

Figure 7. The separation results of membranes when separating single component oil/water mixture. (**a**) TOC (the Total Organic Carbon) value and calculated separation efficiency of the Cu(OH)$_2$-CS membrane. (**b**) TOC (the Total Organic Carbon) value and calculated separation efficiency of the Ag-CS membrane.

For further analysis of the reason for the high efficiency of membranes after 10 instances of separation, Figure 8a,b clearly illustrate the Cu(OH)$_2$-CS and Ag-CS membrane. From Figure 8a,b, it can be clearly seen that no visible generated structures were destroyed. For the same reason, no loss of the rough and feathery micro/nanoscale structure on the membranes contributed to the very high efficiency of the membranes after 10 instances of repeated separation. The good performance shown in the experiment will greatly benefit the application of these membranes in practical usage.

Figure 8. The SEM images of the prepared membranes after repetitive experiment: (**a**) the SEM image of the Cu(OH)$_2$-CS membrane after repetitive experiment; (**b**) the SEM image of the Ag-CS membrane after repetitive experiment.

For the practical application of the prepared separation membranes, the resistance to acid and alkali solutions, intrusion pressure and liquid flux are also of significant importance. Considering the component and pH value of actual oil/water mixtures are complex, which may affect the performance of the membranes in the separation experiment, the acid and alkali resistance of the two membranes were tested to ensure the better application of the prepared membranes in practice. The solutions used for the test are hydrochloric acid (HCl) solution with pH 1 and sodium hydroxide (NaOH) solution with pH 13. These two types of membranes were immersed in the excess acid solution and sodium hydroxide solution for 24 h and then subjected to separation experiments to test their performance. The Cu(OH)$_2$-CS membrane turned bright gold after being immersed in acid solution for 24 h, which indicates that the nanoneedle Cu(OH)$_2$ structure reacted with the acid solution. The Cu(OH)$_2$-CS membrane has no obvious change after immersing in sodium hydroxide solution for 24 h because the preparation process of the nanoneedle Cu(OH)$_2$-CS membrane is in alkali solutions, which indicates that the membrane has a good alkali resistance.

The contact angle test showed that the Cu(OH)$_2$-CS membrane retained superhydrophilic/superlipophilic properties in air and underwater-superoleophobicity after immersion. The results of separation experiments exhibited that Cu(OH)$_2$-CS membrane also can separate the oil/water mixture with high separation efficiency, although the pressure capacity decreases slightly. This is mainly because the nanoneedle Cu(OH)$_2$ structure reacted with excess acid in solution to form CuCl$_2$ dissolved in the solution. However, the substrate was no longer smooth at this time, because the substrate exhibited a similar etching state with micro-nanostructure after the reaction occurred between nanoneedle Cu(OH)$_2$ structure and acid solution. Moreover, the substrate did not continue to react with the acid solution, so the membrane retained its original wettability characteristics and can be used for oil/water separation.

For the generated feathery structure on the Ag-CS membrane, there was no obvious change in the appearance after immersing in an acidic solution and in the alkaline solution. The results of the wettability and separation test showed no significant difference between the membrane before and after immersion. As per the analysis, neither Ag nor the Cu substrate will react in the nonoxidizing strong acid and strong alkali solution, and there was no significant change in wettability and separation performance.

The intrusion pressure determines how much oil the prepared membranes can support, and in the study, we calculated the intrusion pressure by the following equation:

$$\Delta P = \frac{2\gamma}{R} = -\frac{l\gamma \cos \theta_a}{A} = -\frac{2\gamma \cos \theta_a}{\pi r} \tag{4}$$

$$\exp = \rho g h_{\max} \tag{5}$$

In Equation (4), ΔP represents the intrusion pressures, γ represents the surface tension, θ_a is the advancing contact angle, A is the area of the hole of membrane, l is the perimeter of the hole, R is the radius of curvature of the meniscus, r is the radius of the hole. When $\Delta P > 0$, additional pressure is required for oil/water mixture to pass the membrane, and when $\Delta P < 0$, the mixture can pass through the membrane without additional pressure. It can be seen from Equation (4) that the advancing contact angle and the surface tension cannot be changed—only the parameters of the membrane can be changed when the oil type and the component of surface are determined. When the $\cos\theta_a$ of the superoleophobic membrane is negative and the radius of the holes increase, the ΔP will decrease. When the ΔP is too small, a small pressure of oil will make the oil pass through the membrane smoothly and almost no liquid column can be withstood. The column height the membranes can withstand gradually increases with the decrease of the radius of the hole, and when the passing velocity of liquid is greater than the cumulative velocity of the liquid column, separation can be achieved. Therefore, the difference of component and radius between the Cu(OH)$_2$-CS and Ag-CS membranes lead to different intrusion pressures of the membranes.

In Equation (5), ρ represents the density of tested oil type (diesel), g denotes the gravitational acceleration, $h_{\max}$ denotes the maximum height of the diesel that prepared membranes were able to support. In the experiment, water passed through membranes smoothly, which means that the intrusion pressure of membranes toward water was 0 because of its superhydrophilicity. However, the membranes also displayed underwater-superoleophobicity, and this made the oil stay above the membranes. With the increase of the oil height at a particular moment, oil permeated and then passed through the membranes, and the maximum bearable height of the prepared membranes was obtained. The maximum height is determined by considering the average of three measurements for each membrane. The maximum height obtained of the Cu(OH)$_2$-CS and Ag-CS membrane was 17.0 cm and 19.2 cm, respectively, as found using Equation (4); the average intrusion pressure of the Cu(OH)$_2$-CS and Ag-CS membrane were 1445 Pa and 1632 Pa, respectively (Figure 9). The relatively high intrusion pressures demonstrate that the membrane had the ability to support a large amount of oil/water mixture, which means the prepared membranes have capacity to separate large amounts of oil/water mixture.

As mentioned above, the smaller radius of the hole will lead to lower separation rate, which will limit the application of the membrane in practice. To evaluate the liquid flux, the single component oily wastewater was made by adding 15 mL diesel and 135 mL deionized water together, and the liquid flux was calculated by measuring the time that the prepared membranes took to completely separate the oil/water mixture at ambient temperature. Figure 9 clearly shows the liquid flux of the prepared membranes were 3651 Kg/(m^2·h) and 3084 Kg/(m^2·h). In this study, the liquid flux actually refers to the water flux, and in this case, the liquid flux of prepared membranes was greatly influenced by the flow area and wettability of the membranes. We clearly found that there were highly non-uniform structures on the surface of the Ag-CS membrane, as shown in Figure 1c,d. It illustrates that the flow area of the Ag-CS membrane is small because some holes were blocked by the generated rough structures, so the liquid flux of the Ag-CS membrane was much smaller than that of the Cu(OH)$_2$-CS membrane. Figure 5a,b show the underwater oil contact angles of the prepared membranes were 160.3° and 156.7° and the Cu(OH)$_2$-CS membrane is relatively more repellent to oil. The results show that the Cu(OH)$_2$-CS membrane attracted more water droplets as compared with the Ag-CS membrane, and water passed through the membrane more smoothly. Considering the two aspects above, the liquid flux of Cu(OH)$_2$-CS membrane is larger than that of the Ag-CS membrane, and this is consistent with the result of the practical experiment presented in Figure 9.

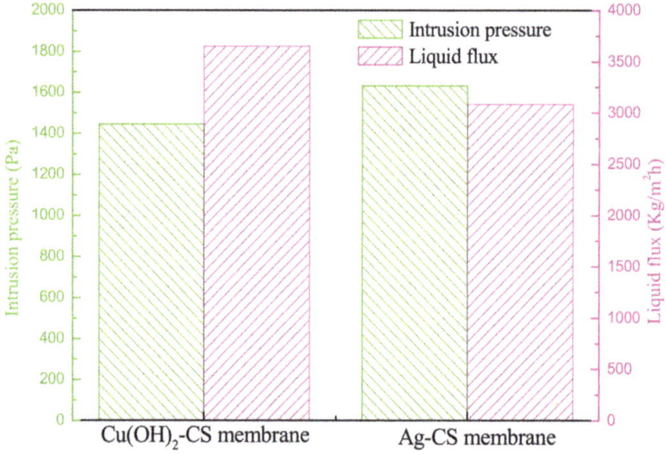

Figure 9. The intrusion pressure and liquid flux of the prepared membranes.

Considering some practical applied conditions mentioned above, including separation efficiency, the resistance to acid and alkali solutions, intrusion pressure and liquid flux, we could find that the Ag-CS membrane showed better performance in the separation of the oil/water mixture compared with the Cu(OH)$_2$-CS membrane and the Ag-CS membrane. These results may have a broad range of application. However, there is a limitation that may restrict the wide range of application of the Ag-CS membrane, which is the relatively higher cost compared with that of the Cu(OH)$_2$-CS membrane. In the further work, we will continue to study how to fabricate lower-cost membranes with similar properties and practical effects, such as obtaining a micro-structured surface via etching methods and then modifying the surface with some lower-priced materials.

3.5. Multicomponent Oil/Water Mixtures Separation

The research was further carried out to the multicomponent oil/water mixture separation in consideration of the practical oily wastewater from factories or living are often contains numerous oil types. Herein, two different types of oils were chosen i.e., sunflower oil and gasoline, to represent macromolecular oils and small molecular oils. So, three different oil/water mixtures were obtained: sunflower oil/water (15 mL + 135 mL), gasoline/water (15 mL + 135 mL) and sunflower oil/gasoline/water (7.5 Ml + 7.5 mL + 135m L). After the oil/water mixtures were prepared, the separation experiments were carried out using the device shown in Figure 6.

Using Equation (3), the separation efficiency of the prepared membranes for multicomponent was evaluated, and the results of TOC value are shown in Figure 10. Figure 10 clearly shows that the prepared membranes all showed notable results for the separation, and the efficiency of the separation was higher than 99.8%. The prepared membranes had even better separation efficiency when the oil/water mixture consisting of macromolecular oil (sunflower oil) was separated as compared with that of small molecular oil (gasoline)/water mixture was separated. Meanwhile, the separation efficiency of different oil types (sunflower oil+ gasoline)/water mixture was somewhere in between. The diameter of suspended oil and dispersed oil droplets was 20–150 um and these oils were mostly the same as prepared for the experiment. The Figure 1c,d clearly show that there were many uniform structures and highly non-uniform structures existed on the surface of the $Cu(OH)_2$-CS membrane and the Ag-CS membrane respectively. It shows that the pore size of the membranes prepared for the experiment was 40 um, i.e., probably larger than that of the gasoline molecular in water, but smaller than that of the sunflower oil. The highly nonuniform structures on the Ag-CS membrane surface further reduced the flow area of the pores. In this case, the prepared membranes likely allowed some small gasoline molecular to pass through but did not allow macro sunflower oil molecules to pass through. Thus, the water sample after separation of sunflower oil/water mixture contained less oil molecules.

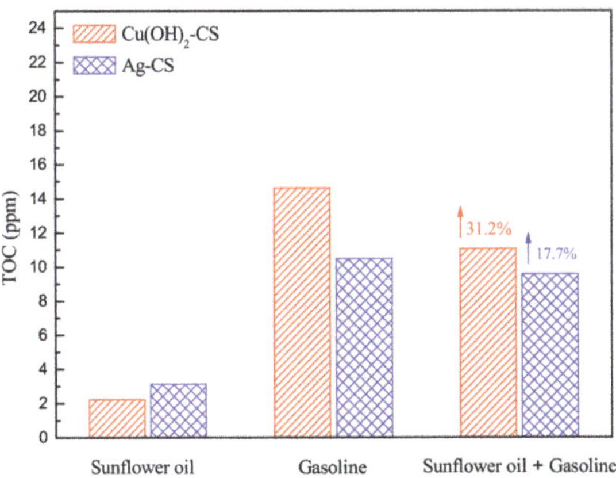

Figure 10. The separation results of prepared membranes for multicomponent oil/water mixture.

By analyzing the TOC data, it was found that the TOC values of prepared membranes are 31.2% and 17.7% higher than the mean values ($TOC_{avg} = \frac{TOC_{sunflower} + TOC_{gasoline}}{2}$) of the two single oils as displayed in Figure 10. Because of excellent mutual solubility of the sunflower oil and gasoline, a few generated oils were naturally made in the prepared oil/water mixture, with sizes between that of the sunflower oil and gasoline droplets. While separating the multicomponent oil/water mixture, some single component oil droplets passed through the membranes, besides, some formed soluble oil droplets whose size was smaller than that of prepared membranes also passed through the membranes. The feathery micro/nanoscale structure generated on the Ag-CS membrane decreased the flow area as compared with that of the $Cu(OH)_2$-CS membrane. In this case, the increased proportion of the TOC value of Ag-CS membrane was smaller than that of the $Cu(OH)_2$-CS membrane.

4. Conclusions

In the current study, focused on the copper substrate, two types of membranes with nanoscale needle structure and the micro/nanoscale feathery structure were successfully made by oxidative reaction and in situ chemical replacement reaction. The prepared membranes were used for systematically separating not only oil/water mixture consisting of single component but also multicomponent mixtures. We also tested the permeation pressure and liquid flux of each prepared membranes and studied the effects of microstructure and surface wettability on these two experimental performances. In comparison, while separating the single component oil/water mixture, it was observed that the separation efficiencies of the prepared membranes were above 99.8%, the Ag-CS membrane exhibited higher efficiency as compared to the $Cu(OH)_2$-CS membrane after 10 repetitions of the experiment. As a result of the combined effect of generated surface structure and surface wettability, the $Cu(OH)_2$-CS membrane showed a higher permeation pressure and supported larger amounts of oil/water mixtures, but in turn, the Ag-CS membrane had higher liquid flux and separated the oil/water mixture more quickly. In the separation for multicomponent oil/water mixture, the prepared membranes all showed good performance with high separation efficiency, and the results of separating multicomponent oil were somehow in between single component oils. The TOC test result of $Cu(OH)_2$-CS membrane and Ag-CS membrane were 31.2% and 17.7% higher than average of the two oils probably because some new oil droplets created by mutual dissolution passed through the membranes. Nevertheless, the prepared membranes still performed well and maintained high efficiencies in separation, thus proving that the prepared membranes, especially the Ag-CS membrane with a feathery micro/nanocomposite structure, could be an exciting application in the field of oil/water separation.

Supplementary Materials: The video about the separation process as the supplementary material is available online at http://www.mdpi.com/1996-1073/12/23/4564/s1.

Author Contributions: The contributions of each author to the research are as follows: Conceptualization, J.Y., C.C., B.Q., J.W., M.A.Q.; Methodology, J.Y., B.Q., J.W., and M.A.Q.; Software, J.Y. and C.C.; Resources, J.Y., and M.A.Q.; Writing-Original Draft Preparation, J.Y.; Writing-Review & Editing, B.Q., J.W. and M.A.Q.

Funding: This research was funded by the project of National Natural Science Foundation of China, China (No. 51776168, No.51636006, No. 51611130060), the Natural Science Foundation of Jiangsu Province, China (No. BK20171236), Funds of International Cooperation and Exchange of the National Natural Science Foundation of China (Research collaboration NSFC-VR) (No. 51961135102), Shaanxi Creative Talents Promotion Plan-Technological Innovation Team (2019TD-039), The Fundamental Research Funds for the Central Universities (Creative Team Plan No.cxtd2017004 in Xi'an Jiaotong University), and the Fundamental Research Funds for the Central Universities, China (No. xjj2017086). The APC was funded by National Natural Science Foundation of China (51776168).

Conflicts of Interest: All authors have read and approved the final submitted manuscript. No conflicts of interest are involved in this work.

References

1. Cheryan, M.; Rajagopalan, N. Membrane processing of oily streams. Wastewater treatment and waste reduction. *J. Membr. Sci.* **1998**, *151*, 13–28. [CrossRef]

2. Padaki, M.; Murali, R.S.; Abdullah, M.S.; Misdan, N.; Moslehyani, A.; Kassim, M.A.; Hilal, N.; Ismail, A.F. Membrane technology enhancement in oil–water separation. A review. *Desalination* **2015**, *357*, 197–207. [CrossRef]

3. Chan, Y.J.; Chong, M.F.; Chung, L.L.; Hassell, D.G. A review on anaerobic-aerobic treatment of industrial and municipal wastewater. *Chem. Eng. J.* **2009**, *155*, 1–18. [CrossRef]

4. Gao, C.; Sun, Z.; Kan, L.; Chen, Y.N.; Cao, Y.Z.; Zhang, S.Y.; Lin, F. Integrated oil separation and water purification by a double-layer TiO_2-based mesh. *Energy Environ. Sci.* **2013**, *6*. [CrossRef]

5. Vemu, S.; Pinnamaneni, U.B. Cross-flow microfiltration of industrial oily wastewater: Experimental and theoretical consideration. *Sep. Sci. Technol.* **2011**, *46*, 1213–1223.

6. Lin, F.; Zhang, Z.Y.; Mai, Z.H.; Liu, B.Q.; Lei, J.; Zhu, D.B. A super-hydrophobic and super-oleophilic coating mesh film for the separation of oil and water. *Angew. Chem. Int. Ed.* **2010**, *43*, 2012–2014.

7. Ivshina, I.B.; Kuyukina, M.S.; Krivoruchko, A.V.; Elkin, A.A.; Makarov, S.O.; Cunningham, C.J.; Peshkur, T.A.; Atlas, R.M.; Philp, J.C. Oil spill problems and sustainable response strategies through new technologies. *Environ. Sci. Process. Impacts* **2015**, *17*, 1201–1219. [CrossRef]

8. Zhang, L.; Wu, J.; Wang, Y.; Long, Y.; Zhao, N.; Xu, J. Combination of bioinspiration: A general route to superhydrophobic particles. *J. Am. Chem. Soc.* **2012**, *134*. [CrossRef]

9. Nguyen, D.D.; Tai, N.H.; Lee, S.B.; Kuo, W.S. Superhydrophobic and superoleophilic properties of graphene-based sponges fabricated using a facile dip coating method. *Energy. Environ. Sci.* **2012**, *7*, 7908–7912. [CrossRef]

10. Jenssen, B.M. Review article: Effects of oil pollution, chemically treated oil, and cleaning on thermal balance of birds. *Environ. Pollut.* **1994**, *86*. [CrossRef]

11. Wang, S.; Song, Y.; Jiang, L. Microscale and nanoscale hierarchical structured mesh films with superhydrophobic and superoleophilic properties induced by long-chain fatty acids. *Nanotechnology* **2007**, *18*, 15103–15105. [CrossRef]

12. Adebajo, M.O.; Frost, R.L.; Kloprogge, J.T.; Carmody, O.; Kokot, S. Porous materials for oil spill cleanup: A review of synthesis and absorbing properties. *J. Porous Mat.* **2003**, *10*, 59–170. [CrossRef]

13. Srijaroonrat, P.; Julien, E.; Aurelle, Y. Unstable secondary oil/water emulsion treatment using ultrafiltration: Fouling control by backflushing. *J. Membr. Sci.* **1999**, *159*, 11–20. [CrossRef]

14. Darling, S.B. Perspective: Interfacial materials at the interface of energy and water. *J. Appl. Phys.* **2018**, *124*. [CrossRef]

15. Zhang, L.B.; Zhang, Z.H.; Peng, W. Smart surfaces with switchable superoleophilicity and superoleophobicity in aqueous media: Toward controllable oil/water separation. *NPG Asia Mater.* **2012**, *4*. [CrossRef]

16. Crick, C.R.; Gibbins, J.A.; Parkin, I.P. Superhydrophobic polymer-coated copper-mesh; membranes for highly efficient oil-water separation. *J. Mater. Chem. A* **2013**, *1*. [CrossRef]

17. Kammerer, M.; Mastain, O.; Dréan-Quenech'Du, S.L.; Pouliquen, H.; Larhantec, M. Liver and kidney concentrations of vanadium in oiled seabirds after the Erika wreck. *Sci. Total Environ.* **2004**, *333*, 295–301. [CrossRef]

18. Kajitvichyanukul, P.; Hung, Y.T.; Wang, L.K. *Oil Water Separation. Handbook of Environmental Engineering*; Wang, L.K., Hung, Y.-T., Shammas, N.K., Eds.; Springer: Berlin, Germany, 2006.

19. Kwon, W.T.; Park, K.; Han, S.D.; Yoon, S.M.; Kim, J.Y.; Bae, W.; Rhee, Y.W. Investigation of water separation from water-in-oil emulsion using electric field. *J. Ind. Eng. Chem.* **2010**, *16*, 684–687. [CrossRef]

20. Eow, J.S.; Ghadiri, M.; Sharif, A.O. Electrostatic and hydrodynamic separation of aqueous drops in a flowing viscous oil. *Chem. Eng. Process. Process Intensif.* **2002**, *41*, 649–657. [CrossRef]

21. Hanafy, M.; Nabih, H.I. Treatment of oily wastewater using dissolved air flotation technique. *Energy Sources* **2007**, *29*, 143–159. [CrossRef]

22. Yong, J.; Chen, F.; Yang, Q.; Bian, H.; Du, G.Q.; Shan, C.; Huo, J.L.; Fang, Y.; Hou, H. Oil-Water Separation: Oil-Water Separation: A Gift from the Desert. *Adv. Mater. Interfaces* **2016**, *3*. [CrossRef]

23. Yang, J.; Zhang, Z.Z.; Xu, X.H.; Zhu, X.T.; Men, X.H.; Zhou, X.Y. Superhydrophilic–superoleophobic coatings. *J. Mater. Chem.* **2012**, *22*, 2834–2837. [CrossRef]

24. Zhang, F.; Zhang, W.B.; Shi, Z.; Wang, D.; Jin, J. Nanowire-Haired inorganic membranes with superhydrophilicity and underwater ultralow adhesive superoleophobicity for high-efficiency oil/water separation. *Adv. Mater.* **2013**, *25*, 4192–4198. [CrossRef]

25. Xue, Z.X.; Liu, M.J.; Jiang, L. Recent developments in polymeric superoleophobic surfaces. *J. Polym. Sci. Part B Polym. Phys.* **2012**, *50*, 1209–1224. [CrossRef]

26. Sun, T.L.; Lin, F.; Gao, X.F.; Jiang, L. Bioinspired surfaces with special wettability. *Acc. Chem. Res.* **2005**, *38*, 644–652. [CrossRef]

27. Lin, F.; Zhang, Y.N.; Xi, J.M.; Zhu, Y.; Wang, N.; Xia, F.; Jiang, L. Petal effect: A superhydrophobic state with high adhesive force. *Langmuir* **2008**, *24*, 4114–4119.

28. Feng, X.J.; Jiang, L. Design and creation of superwetting/antiwetting surfaces. *Adv. Mat.* **2010**, *18*, 3063–3078. [CrossRef]

29. Zhang, Y.L.; Xia, H.; Kim, E.Y.; Sun, H.B. Recent developments in superhydrophobic surfaces with unique structural and functional properties. *Soft Matter* **2012**, *8*, 11217. [CrossRef]

30. Blossey, R. Self-cleaning surfaces—Virtual realities. *Nat. Mat.* **2003**, *2*, 301–306. [CrossRef]

31. Zhang, J.; Seeger, S. Superoleophobic coatings with ultralow sliding angles based on silicone nanofilaments. *Angew. Chem. Inte. Ed.* **2011**, *50*, 6652–6656. [CrossRef]

32. Roach, P.; Shirtcliffe, N.J.; Newton, M.I. Progess in superhydrophobic surface development. *Soft Matter* **2008**, *4*, 224–240. [CrossRef]

33. Lafuma, A.; Quéré, D. Superhydrophobic states. *Nature Mater.* **2003**, *2*, 457–460. [CrossRef] [PubMed]

34. Zhang, X.; Shi, F.; Niu, J.; Jiang, Z.Q. Superhydrophobic surfaces: From structural control to functional application. *J. Mater. Chem.* **2008**, *18*, 621–633. [CrossRef]

35. Cheng, M.J.; Gao, Y.F.; Guo, X.P.; Shi, Z.Y.; Chen, J.F.; Shi, F. A functionally integrated device for effective and facile oil spill cleanup. *Langmuir* **2011**, *27*, 7371–7375. [CrossRef] [PubMed]

36. Erbil, H.Y.; Demirel, A.L.; Avci, Y.; Mert, O. Transformation of a simple plastic into a superhydrophobic surface. *Science* **2003**, *299*, 1377–1380. [CrossRef] [PubMed]

37. Xue, Z.X.; Sun, Z.X.; Cao, Y.Z.; Chen, Y.N.; Tao, L.; Li, K.; Lin, F.; Fu, Q.; Wen, Y. Superoleophilic and superhydrophobic biodegradable material with porous structures for oil absorption and oil–water separation. *RSC Adv.* **2013**, *3*, 23432–23437. [CrossRef]

38. Zhang, J.L.; Huang, W.H.; Han, Y.C. A composite polymer film with both superhydrophobicity and superoleophilicity. *Macromol. Rapid Commun.* **2010**, *27*, 804–808. [CrossRef]

39. Tu, C.W.; Tsai, C.H.; Wang, C.F.; Kuo, S.W.; Chang, F.C. Fabrication of superhydrophobic and superoleophilic polystyrene surfaces by a facile one-step method. *Macromol. Rapid Commun.* **2010**, *28*, 2262–2266. [CrossRef]

40. Choi, H.O.; Zhang, K.; Dionysiou, D.D. Effect of permeate flux and tangential flow on membrane fouling for wastewater treatment. *Sep. Purif. Technol.* **2015**, *45*, 68–78. [CrossRef]

41. Huang, Y.; Zha, G.Y.; Luo, Q.J.; Zhang, J.X.; Zhang, F.; Li, X.H.; Zhao, S.F.; Zhu, W.P.; Li, X.D. The construction of hierarchical structure on Ti substrate with superior osteogenic activity and intrinsic antibacterial capability. *Sci. Rep.* **2014**, *4*, 6172. [CrossRef]

42. Zheng, J.F.; He, A.H.; Li, J.X.; Han, C.C. Studies on the controlled morphology and wettability of polystyrene surfaces by electrospinning or electrospraying. *Polymer* **2006**, *47*, 7095–7102. [CrossRef]

43. Wang, B.; Liang, W.X.; Guo, Z.G.; Liu, W.M. Biomimetic super-lyophobic and super-lyophilic materials applied for oil/water separation: A new strategy beyond nature. *Chem. Soc. Rev.* **2015**, *44*, 336–361. [CrossRef] [PubMed]

44. Yang, H.C.; Xie, Y.S.; Chan, H.; Narayanan, B.; Chen, L.; Waldman, R.Z.; Sankaranarayanan, S.K.R.S.; Elam, J.W.; Darling, S.B. Crude-oil-repellent membranes by atomic layer deposition: Oxide interface engineering. *ACS Nano* **2018**, *12*, 8678–8685. [CrossRef] [PubMed]

45. Kota, A.K.; Kwon, G.B.; Choi, W.J. Hygro-responsive membranes for effective oil–water separation. *Nat. Commun.* **2012**, *3*, 1025. [CrossRef]

46. Zhu, X.T.; Zhang, Z.Z.; Men, X.H.; Yang, J.; Wang, K.; Xu, X.H.; Zhou, X.Y.; Xue, Q.J. Robust superhydrophobic surfaces with mechanical durability and easy repairability. *J. Mater. Chem.* **2011**, *21*, 15793–15797. [CrossRef]

47. Wenzel, R.N. Resistance of solid surfaces to wetting by water. *Ind. Eng. Chem.* **1936**, *28*, 988–994. [CrossRef]

Article

Evolution of Virtual Water Transfers in China's Provincial Grids and Its Driving Analysis

Yiyi Zhang, Shengren Hou, Jiefeng Liu *, Hanbo Zheng, Jiaqi Wang and Chaohai Zhang

Guangxi Key Laboratory of Power System Optimization and Energy Technology, Guangxi University, Nanning 530004, Guangxi, China; yiyizhang@gxu.edu.cn (Y.Z.); 15670131688@163.com (S.H.); hanbozheng@163.com (H.Z.); wangjiaqijiaqi@gmail.com (J.W.); zch852@126.com (C.Z.)
* Correspondence: liujiefeng2018@gxu.edu.cn; Tel.: +86-199-6812-0257

Received: 19 November 2019; Accepted: 16 December 2019; Published: 9 January 2020

Abstract: In China, electricity transmission has increased rapidly over the past decades, and a large amount of virtual water is delivered from power generation provinces to load hubs. Understanding the evolution of the virtual water network embodied in electricity transmission is vital for mitigating water scarcity. However, previous studies mainly calculated the virtual water transferred in short periods in low-spatial resolution and failed to reveal driving forces of the evolution of virtual water. To solve this problem, we investigated the historical evolution of the virtual water network and virtual scarce water network embodied in interprovincial electricity transmission between 2005 and 2014. The driving forces of the evolution of virtual (scarce) water networks were analyzed at both national level and provincial level. The results show that the overall virtual water transmission and virtual scarce water transmission increased by five times, and the direction was mainly from southwest and northwest provinces to eastern provinces. Sichuan, Yunnan, and Guizhou played an increasingly important role in virtual water exporting, and northwestern provinces had dominated the virtual scarce water exporting in the decade. At the national level, the increase of virtual water is mainly driven by the change of power generation mix and power transmission. At the provincial level, the increase of virtual water transmission in the largest virtual water exporter (Sichuan) is driven by the power generation mix and the power transmission, between 2005 and 2010, and 2010 and 2014, respectively. Considering the expanding of electricity transmission, the development of hydropower in the southwestern provinces and other renewable energies (solar and wind) in the northeastern provinces would overall mitigate the water scarcity in China.

Keywords: virtual water network; inter-provincial electricity transmission; structural decomposition analysis; electricity-water nexus

---

1. Introduction

Water scarcity is serious in China, and per capita water resources are only 25% of the average world level [1]. The spatial and temporal distribution of water resources is extremely unbalanced [2]. Over 80% of water resources are reserved in southern China, while northern China sustains 47% of people with less than 20% of water resources. Water scarcity has become a major bottleneck restricting the sustainable development of the economy and society [3]. The Chinese government has implemented the strictest regulation to mitigate water scarcity, e.g., constraining the total use of water resources within 700 billion cubic meters by 2030 [4]. Industry sectors consumed 126 billion cubic meter water accounting for 20% of the total water consumption in 2018 [5]. Among all industrial sectors, the power system is the largest water consumer contributing to 70% of the total industrial water consumption [6]. Therefore, investigating electricity-related water consumption is a key aspect to mitigate water scarcity in China.

To investigate water stress posed by the power system, previous studies have quantified direct water use or consumption of power generation in China [7,8], focusing on different power generation technologies [9,10], environmental impacts [11,12] and future scenarios analysis [13,14]. To further quantify the life cycle water consumption of functional electricity (kWh), virtual water (VW) was introduced in the investigation for electricity-water nexus. The concepts of virtual water and water footprint were proposed to quantify water consumption through the life cycle of a commodity [15,16]. The life cycle water use of power generation was estimated through different methods [17,18] or at different spatial scales [19–21]. For instance, Feng et al. [17] used the hydro-life-cycle-analysis method to calculate water coefficients of eight power generation types, and they found the water coefficient of hydropower is higher than that of thermal power. Besides, electricity is transferred from arid western provinces to eastern provinces. The mismatch between energy resources and water resources in China has raised many concerns regarding potential water-related impacts from power transmission.

According to the virtual water strategy [16], VW flows are transferred from the source to the sink and thus reallocates the water resources on both sides. In the power system, VW is delivered from power generation provinces to load hubs through electricity transmission, which may induce water scarcity. Multiple studies have explored the virtual water network through electricity transmission in recent years. For instance, Zhu et al. [20] and Guo et al. [21] developed a node-flow model to calculate the virtual water transmission (VWT) among six sub-national grids in the year 2010 and short-time span (2007–2012), respectively. Zhang et al. [22] quantified the interprovincial VWT embodied in thermal power transmission and adjusted VW into virtual scarce water (VSW) in 2011, by improving the spatial resolution to the provincial level and combining the water stress index (WSI). The aforementioned studies proved that VWT extremely exacerbated the local water scarcity in power generation provinces, especially in northwestern provinces. In USA, Chini et al. [23] explored the blue water and grey water transfers embodied in power grids and provided clear grey water and blue water network. In addition, some studies identified whether the virtual water flows mitigate or exacerbate the water scarcity by comparing the WSI of source and sink [24]. However, there is a lack of study for the evolution of VWT embodied in electricity transmission.

Electricity transmission has increased by 4.2 times in the last decade [25], resulting in a significant increase of VWT. To explore the evolution of VWT, the decomposition analysis is introduced to reveal the driving forces of the evolution of the virtual water network. The decomposition analysis was proposed to quantify different contributions of factors to the change of an indicator between two periods. Structural decomposition analysis (SDA) and index decomposition analysis (IDA) are two widely used approaches to investigate electricity consumption related indicators, such as carbon emissions [26,27] and virtual water consumption [28,29]. Logarithmic Mean Divisia Index (LMDI) is the most popular IDA method with low-resolution data requirements and simplicity. It could also be used to investigate electricity-related carbon emission [30] and water consumption [31]. However, unlike SDA, LMDI cannot distinguish final demand and intermediate demand, thus, the indirect impacts of change in final demand cannot be estimated. To estimate the contribution of indirect impacts to the overall changes of energy-related indicators, the SDA method was widely used. Nevertheless, SDA is restricted by its high-quality data requirements based on the input-output (IO) table. Research [32] proposed a modified SDA based on the electricity transmission table to analyze the contribution of factors (especially for the change of power transmission) to the increase of carbon emission in the power sector. Electricity-related water footprint differs from carbon footprint because water resources vary from region to region when carbon emissions have the same effects on climate change regardless of location. Paper [33] investigated the driving forces for the evolution of VWT embodied in thermal power generation, and the results showed that water efficiency improvements were the main driver to the decrease of VWT at the national level. However, the analysis in paper [33] is not comprehensive, because only thermal power was considered, and the differences of driving forces in various provinces were not included. During the last decades, China's power generation and transmission have expanded significantly, which was largely driven by renewable energy technologies.

Hydropower has a higher water coefficient than thermal power because of evaporation in reservoir. A deep investigation of the evolution of virtual water transfers is necessary for the understanding of water-electricity nexus.

This study investigated the long-time series evolution of virtual (scarce) water network embodied in interprovincial electricity trade, and further identified the driving factors to the change of the virtual (scarce) water transmission by using a modified SDA model. Compared to previous studies, the novelty and significance of this work are as follows: (1) water consumption for all power generation technologies (including thermal, hydropower, wind, solar, and nuclear) are included; (2) VW network and VSW network are both investigated to represent the imbalance spatial distribution of water resources in different provinces; and (3) by using a modified SDA model, a high-resolution decomposition results are analyzed to capture the contribution of different factors at both national level and provincial level.

The structure of this study is into two parts. In part 1, Sections 2.1 and 2.2 constructed the node-flow model of the virtual water network, and the SDA model for investigating the driving factors, respectively. Sections 2.3 and 2.4 introduced the spatial and temporal area of the paper and the data sources. In part 2, Section 3 shows the results for the VW network evolution and its driving factors. Sections 4.1 and 4.2 discussed the impacts of different policies between 2005 to 2014 and advice for Chinese policymaker, while Section 4.3 discussed the limitation and advantages in this study.

2. Materials and Methods

2.1. Modeling Virtual Water Network Embodied in Electricity Trade

Based on the node-flow model proposed in previous studies [20,21], we modelled inter-provincial electricity transmission as a network with nodes and flows. Each province and each electricity trade is assumed as a node and a flow, respectively. For each node, the total power demand is supplied by local power generation and power imported from other nodes; local plants can only outflow power to satisfy the local demand and other nodes. The transmission loss is ignored in this study, and the balance between power generation and power demand for each node can be expressed below:

$$\sum_{i=m}^{i=1} ED_i + \sum_{i=m,i\neq j}^{i=1} ET_{ij} = \sum_{i=m}^{i=1} EG_i + \sum_{j=m,j\neq i}^{j=1} ET_{ji} \tag{1}$$

where EG_i is the electric power generation of node i, ED_i is the power demand of node i, and ET_{ij} is the electricity flow from node i to node j.

The balance between power demand and power generation can be adapted to the virtual water transfers by Equation (2):

$$\sum_{i=m}^{i=1} EG_i{\cdot}dw_i + \sum_{i=m,i\neq j}^{i=1} ET_{ij}{\cdot}tw_j = \left(\sum_{i=m}^{i=1} ED_i + \sum_{j=m,j\neq i}^{j=1} ET_{ji}\right){\cdot}tw_i \tag{2}$$

where dw_i represents the direct water consumption factor for functional unit power generation in province i, and tw_j represents the embodied water footprint (total water consumption factor) of functional unit power supplied by province j.

We assume that all electricity feeding into province i is totally mixed, i.e., from different sources. The embodied water footprint of electricity supplied to local users is thus the same as the electricity exported to other nodes for provinces i. To calculate twi, Equation (2) can be rearranged into Equations (3) and (4):

$$\sum_{i=m}^{i=1} EG_i{\cdot}dw_i = \left(\sum_{i=m}^{i=1} ED_i + \sum_{j=m,j\neq i}^{j=1} ET_{ji}\right){\cdot}tw_i - \sum_{i=m,i\neq j}^{i=1} ET_{ij}{\cdot}tw_j \tag{3}$$

$$\hat{E}G \cdot DW = M \cdot TW \tag{4}$$

where $\hat{E}G$ represents the diagonal matrix of electricity generation EG_i, DW represents the column of dw_i, TW represents the column of tw_i, and M is rearranged as electricity trade matrix given by Equation (5):

$$M = \begin{bmatrix} \sum_{j \neq 1} ET_{1j} + E_{d,1} & -ET_{21} & \cdots & -ET_{m1} \\ -ET_{12} & \sum_{j \neq 2} ET_{2j} + E_{d,2} & \cdots & -ET_{m2} \\ \vdots & \vdots & \ddots & \vdots \\ -ET_{1m} & -ET_{2m} & \cdots & \sum_{j \neq m} ET_{2j} + E_{d,2} \end{bmatrix} \tag{5}$$

The vector of total water coefficient can be calculated in the following:

$$TW = M^{-1} \cdot \hat{E}D \cdot DW = H \cdot DW \tag{6}$$

By using Equation (6), we create a linear map between the direct water coefficient and the virtual water coefficient, thus link the water footprint from the electricity consumption side to the production side. Each element in matrix H, h_{ij}, represents the complete electricity transmission, which is different from the direct electricity transmission, because it includes higher-order electricity transmission. The role of H is the same as the role of Leonfief matrix in the input–output analysis which transforms the direct consumption to the total consumption.

2.2. Virtual Water Transmission and Decomposition Analysis Model

SDA can estimate the contribution of different factors to the overall evolution of an indicator, and it has been widely used to investigate the energy-related emissions. However, SDA is highly depended on the IO table, which restricts the application of itself. According to a previous study [32], a modified SDA model applying the electricity transmission table (not IO table) is used in our study. Combing the power generation and demand of each province, we develop the power transmission table as follows:

$$AT = \left\{AT_{ij}\right\}_{m \times m}, AT_{ij} = ET_{ij}/ED_j \tag{7}$$

$$BT = \left\{BT_{ij}\right\}_{m \times m}, BT_{ij} = ET_{ij}/EG_i \tag{8}$$

where AT_{ji} in matrix AT is the power generation in i_{th} province feed into j_{th} province's functional unit consumption, thus AT can be expressed as the power demand structure. BT_{ij} is the number of power outflows to j_{th} provinces in i_{th} provinces' power generation, which can be regarded as the power generation structure.

Therefore, matrix AT can be regarded as a power demand table and matrix BT should be considered as the power generation structure table. In this case, the total power outflow of a province can be represented as:

$$E_{out} = AT \cdot ED \tag{9}$$

where ED refers to the column for power demand in all provinces and electricity inflow of each province can be represented as:

$$E_{in} = BT^T \cdot EG \tag{10}$$

Therefore, the electricity generation can be represented as:

$$EG = E_{out} - E_{in} + ED = AT \cdot ED - BT^T \cdot EG + ED \tag{11}$$

$$EG = \left(I + BT^T\right)^{-1} \cdot (I + AT) \cdot ED \tag{12}$$

The *VWT* embodied in electricity transmission can be calculated as:

$$VWT = TW^T \cdot E_{in} = TW^T \cdot BT^T \cdot EG = DW^T \cdot H^T \cdot BT^T \cdot \left(I + BT^T\right)^{-1} \cdot (I + AT) \cdot ED \tag{13}$$

where TW^T is the transpose of TW, and I refers to the identity matrix. To classify the driving factor of *VWT*, Equation (13) could be further transformed into Equation (14):

$$VWT = DW^T \cdot H^T \cdot PG \cdot PD \cdot ED \tag{14}$$

where DW is the direct water coefficient, which is dominated by the power generation mix of each province, and H is the revised power transmission structure.

Matrix DW can be regarded as the power generation mix, while H can be regarded as a power transmission structure, and matrix ED can be regarded as the power demand factor. As is described in the previous study [32], at the national level, factor PG and PD can be expressed as power generation structure and power demand structure. At the province level, it should be considered as the external power generation structure and the external power demand structure.

Compared to the SDA model using the input–output Table, this modified model uses electricity transmission data, and could introduce power generation structure and power demand structure to the VWT change. The complete additive decomposition method can eliminate the residual of decomposition [32,34], Equation (15) is thus expressed as an example to decompose the VWT between two periods.

$$\begin{aligned} \Delta VWT_t &= VWT_{t+1} - VWT_t \\ &= E(\Delta DW) + E(\Delta H) + E(\Delta PD) + E(\Delta PG) + E(\Delta ED) \end{aligned} \tag{15}$$

where $E(\Delta DW)$ refers to the contribution of the power generation mix, $E(\Delta H)$ refers to the contribution of the electricity transmission structure, $E(\Delta PD)$ refers to the contribution of the (external) power demand structure, $E(\Delta PG)$ refers to the contribution of the (external) power generation structure, and $E(\Delta ED)$ refers to the contribution of the power demand.

In an SDA model, to obtain a complete decomposition form, each factor should be weighted by Laspeyres or Paasche weights [27]. However, SDA would produce N! different decomposition forms when it has N factors. The N decomposition forms represent N ways to eliminate residuals. In previous studies [28,32], to reduce the computational complexity, residuals are divided equally among factors by using the average residuals for different decomposition forms. This study also uses the averages of different forms to eliminate residuals.

The decomposition formulation of modified SDA model between the two periods can be expressed as follows:

$$E\left(\Delta DW^T\right) = \frac{1}{5} \left(\begin{array}{l} \Delta DW^T \cdot H_t \cdot PG_t \cdot PD_t \cdot EG + \Delta DW^T \cdot H_{t+1} \cdot PG_{t+1} \cdot PD_{t+1} \cdot EG_{t+1} \\[4pt] + \frac{1}{20} \left(\begin{array}{l} \Delta DW^T \cdot H_{t+1} \cdot PG_t \cdot PD_t \cdot EG_t + \Delta DW^T \cdot H_t \cdot PG_{t+1} \cdot PD_t \cdot EG_t + \\ \Delta DW^T \cdot H_t \cdot PG_t \cdot PD_{t+1} \cdot EG_t + \Delta DW^T \cdot H_t \cdot PG_t \cdot PD_t \cdot EG_{t+1} + \\ \Delta DW^T \cdot H_t \cdot PG_{t+1} \cdot PD_{t+1} \cdot EG_{t+1} + \Delta DW^T \cdot H_{t+1} \cdot PG_t \cdot PD_{t+1} \cdot EG_{t+1} + \\ \Delta DW^T \cdot H_{t+1} \cdot PG_{t+1} \cdot PD_t \cdot EG_{t+1} + \Delta DW^T \cdot H_{t+1} \cdot PG_{t+1} \cdot PD_{t+1} \cdot EG_t \end{array} \right) \\[4pt] + \frac{1}{30} \left(\begin{array}{l} \Delta DW^T \cdot H_{t+1} \cdot PG_{t+1} \cdot PD_t \cdot EG_t + \Delta DW^T \cdot H_{t+1} \cdot PG_t \cdot PD_{t+1} \cdot EG_t + \\ \Delta DW^T \cdot H_{t+1} \cdot PG_t \cdot PD_t \cdot EG_{t+1} + \Delta DW^T \cdot H_t \cdot PG_{t+1} \cdot PD_{t+1} \cdot EG_t + \\ \Delta DW^T \cdot H_t \cdot PG_{t+1} \cdot PD_t \cdot EG_{t+1} + \Delta DW^T \cdot H_t \cdot PG_t \cdot PD_{t+1} \cdot EG_{t+1} \end{array} \right) \end{array} \right) \tag{16}$$

$$E\left(\Delta H^T\right) = \frac{1}{5}\left(\begin{array}{l} DW_t^T \cdot \Delta H^T \cdot PG_t \cdot PD_t \cdot EG_t + DW_{t+1}^T \cdot \Delta H^T \cdot PG_{t+1} \cdot PD_{t+1} \cdot EG_{t+1} \\[4pt] + \frac{1}{20}\left(\begin{array}{l} DW_{t+1}^T \cdot \Delta H^T \cdot PG_t \cdot PD_t \cdot EG_t + DW_t^T \cdot \Delta H^T \cdot PG_{t+1} \cdot PD_t \cdot EG_t + \\ DW_t^T \cdot \Delta H^T \cdot PG_t \cdot PD_{t+1} \cdot EG_t + DW_t^T \cdot \Delta H^T \cdot PG_t \cdot PD_t \cdot EG_{t+1} + \\ DW_t^T \cdot \Delta H^T \cdot PG_{t+1} \cdot PD_{t+1} \cdot EG_{t+1} + DW_{t+1}^T \cdot \Delta H^T \cdot PG_t \cdot PD_{t+1} \cdot EG_{t+1} + \\ DW_{t+1}^T \cdot \Delta H^T \cdot PG_{t+1} \cdot PD_t \cdot EG_{t+1} + DW_{t+1}^T \cdot \Delta H^T \cdot PG_{t+1} \cdot PD_{t+1} \cdot EG_t \end{array}\right) \\[4pt] + \frac{1}{30}\left(\begin{array}{l} DW_{t+1}^T \cdot \Delta H^T \cdot PG_{t+1} \cdot PD_t \cdot EG_t + DW_{t+1}^T \cdot \Delta H^T \cdot PG_t \cdot PD_{t+1} \cdot EG_t + \\ DW_{t+1}^T \cdot \Delta H^T \cdot PG_t \cdot PD_t \cdot EG_{t+1} + DW_t^T \cdot \Delta H^T \cdot PG_{t+1} \cdot PD_{t+1} \cdot EG_t + \\ DW_t^T \cdot \Delta H^T \cdot PG_{t+1} \cdot PD_t \cdot EG_{t+1} + DW_t^T \cdot \Delta H^T \cdot PG_t \cdot PD_{t+1} \cdot EG_{t+1} \end{array}\right) \end{array}\right) \tag{17}$$

$$E(\Delta PG) = \frac{1}{5}\left(\begin{array}{l} DW_t^T \cdot H_t^T \cdot \Delta PG \cdot PD_t \cdot EG_t + DW_{t+1}^T \cdot H_{t+1}^T \cdot \Delta PG \cdot PD_{t+1} \cdot EG_{t+1} \\[4pt] + \frac{1}{20}\left(\begin{array}{l} DW_{t+1}^T \cdot H_t^T \cdot \Delta PG \cdot PD_t \cdot EG_t + DW_t^T \cdot H_{t+1}^T \cdot \Delta PG \cdot PD_t \cdot EG_t + \\ DW_t^T \cdot H_t^T \cdot \Delta PG \cdot PD_{t+1} \cdot EG_t + DW_t^T \cdot H_t^T \cdot \Delta PG \cdot PD_t \cdot EG_{t+1} + \\ DW_t^T \cdot H_{t+1}^T \cdot \Delta PG \cdot PD_{t+1} \cdot EG_{t+1} + DW_{t+1}^T \cdot H_t^T \cdot \Delta PG \cdot PD_{t+1} \cdot EG_{t+1} + \\ DW_{t+1}^T \cdot H_{t+1}^T \cdot \Delta PG \cdot PD_t \cdot EG_{t+1} + DW_{t+1}^T \cdot H_{t+1}^T \cdot \Delta PG \cdot PD_{t+1} \cdot EG_t \end{array}\right) \\[4pt] + \frac{1}{30}\left(\begin{array}{l} DW_{t+1}^T \cdot H_{t+1}^T \cdot \Delta PG \cdot PD_t \cdot EG_t + DW_{t+1}^T \cdot H_t^T \cdot \Delta PG \cdot PD_{t+1} \cdot EG_t + \\ DW_{t+1}^T \cdot H_t^T \cdot \Delta PG \cdot PD_t \cdot EG_{t+1} + DW_t^T \cdot H_{t+1}^T \cdot \Delta PG \cdot PD_{t+1} \cdot EG_t + \\ DW_t^T \cdot H_{t+1}^T \cdot \Delta PG \cdot PD_t \cdot EG_{t+1} + DW_t^T \cdot H_t^T \cdot \Delta PG \cdot PD_{t+1} \cdot EG_{t+1} \end{array}\right) \end{array}\right) \tag{18}$$

$$E(\Delta PD) = \frac{1}{5}\left(\begin{array}{l} DW_t^T \cdot H_t^T \cdot PG_t \cdot \Delta PD \cdot EG_t + DW_{t+1}^T \cdot H_{t+1}^T \cdot PG_{t+1} \cdot \Delta PD \cdot EG_{t+1} \\[4pt] + \frac{1}{20}\left(\begin{array}{l} DW_{t+1}^T \cdot H_t^T \cdot PG_t \cdot \Delta PD \cdot EG_t + DW_t^T \cdot H_{t+1}^T \cdot PG_t \cdot \Delta PD \cdot EG_t + \\ DW_t^T \cdot H_t^T \cdot PG_{t+1} \cdot \Delta PD \cdot EG_t + DW_t^T \cdot H_t^T \cdot PG_t \cdot \Delta PD \cdot EG_{t+1} + \\ DW_t^T \cdot H_{t+1}^T \cdot PG_{t+1} \cdot \Delta PD \cdot EG_{t+1} + DW_{t+1}^T \cdot H_t^T \cdot PG_{t+1} \cdot \Delta PD \cdot EG_{t+1} + \\ DW_{t+1}^T \cdot H_{t+1}^T \cdot PG_t \cdot \Delta PD \cdot EG_{t+1} + DW_{t+1}^T \cdot H_{t+1}^T \cdot PG_{t+1} \cdot \Delta PD \cdot EG_t \end{array}\right) \\[4pt] + \frac{1}{30}\left(\begin{array}{l} DW_{t+1}^T \cdot H_{t+1}^T \cdot PG_t \cdot \Delta PD \cdot EG_t + DW_{t+1}^T \cdot H_t^T \cdot PG_{t+1} \cdot \Delta PD \cdot EG_t + \\ DW_{t+1}^T \cdot H_t^T \cdot PG_t \cdot \Delta PD \cdot EG_{t+1} + DW_t^T \cdot H_{t+1}^T \cdot PG_{t+1} \cdot \Delta PD \cdot EG_t + \\ DW_t^T \cdot H_{t+1}^T \cdot PG_t \cdot \Delta PD \cdot EG_{t+1} + DW_t^T \cdot H_t^T \cdot PG_{t+1} \cdot \Delta PD \cdot EG_{t+1} \end{array}\right) \end{array}\right) \tag{19}$$

$$E(\Delta EG) = \frac{1}{5}\left(\begin{array}{l} DW_t^T \cdot H_t^T \cdot PG_t \cdot PD_t \cdot \Delta EG + DW_{t+1}^T \cdot H_{t+1}^T \cdot PG_{t+1} \cdot PD_{t+1} \cdot \Delta EG \\[4pt] + \frac{1}{20}\left(\begin{array}{l} DW_{t+1}^T \cdot H_t^T \cdot PG_t \cdot PD_t \cdot \Delta EG + DW_t^T \cdot H_{t+1}^T \cdot PG_t \cdot PD_t \cdot \Delta EG + \\ DW_t^T \cdot H_t^T \cdot PG_{t+1} \cdot PD_t \cdot \Delta EG + DW_t^T \cdot H_t^T \cdot PG_t \cdot PD_{t+1} \cdot \Delta EG + \\ DW_t^T \cdot H_{t+1}^T \cdot PG_{t+1} \cdot PD_{t+1} \cdot \Delta EG + DW_{t+1}^T \cdot H_t^T \cdot PG_{t+1} \cdot PD_{t+1} \cdot \Delta EG + \\ DW_{t+1}^T \cdot H_{t+1}^T \cdot PG_t \cdot PD_{t+1} \cdot \Delta EG + DW_{t+1}^T \cdot H_{t+1}^T \cdot PG_{t+1} \cdot PD_t \cdot \Delta EG \end{array}\right) \\[4pt] + \frac{1}{30}\left(\begin{array}{l} DW_{t+1}^T \cdot H_{t+1}^T \cdot PG_t \cdot PD_t \cdot \Delta EG + DW_{t+1}^T \cdot H_t^T \cdot PG_{t+1} \cdot PD_t \cdot \Delta EG + \\ DW_{t+1}^T \cdot H_t^T \cdot PG_t \cdot PD_{t+1} \cdot \Delta EG + DW_t^T \cdot H_{t+1}^T \cdot PG_{t+1} \cdot PD_t \cdot \Delta EG + \\ DW_t^T \cdot H_{t+1}^T \cdot PG_t \cdot PD_{t+1} \cdot \Delta EG + DW_t^T \cdot H_t^T \cdot PG_{t+1} \cdot PD_{t+1} \cdot \Delta EG \end{array}\right) \end{array}\right) \tag{20}$$

As we can see, even though the average has been used in this model, the structure of SDA is still very complex.

2.3. Study Area

The spatial area of this study is China's provincial power grids, and the time span is 2005 to 2014. China's power grid is the largest artificial grid in the world, thus choosing it as an example can bring us more details than other countries. From 2005 to 2014, China's power grid has expanded rapidly since the significant development of the economy. The electricity consumption of China increased from 2323 TWh to 5328 TWh from 2005 to 2014. China has a significant imbalance spatial distribution between energy sources and developed level, i.e., most energy sources are distributed far from load hubs. For example, coal-based thermal plants mostly located in northwestern provinces, while hydropower plants are located in southwestern provinces. However, over half of the electricity is consumed by coastal developed regions i.e., Jing-Jin-Ji area, Yangtze River delta, and Guangdong. A large amount of electricity in China is thus generated in the southwestern and northwestern China and then delivered to the load hubs through transmission lines. Power transmission demand is increasing rapidly in the decade, and the maturity of technologies and materials has encouraged the construction of ultra-high-voltage (UHV) transmission projects

Multi-policies have been implemented during the decade, which jointly affected the VWT network. From 2005 to 2010, a large amount of coal power plants were encouraged to construct in western provinces, including Shanxi, Shaanxi, Inner Mongolia, Guizhou, and Yunnan [35]. Then, according to the 12th Five-Year plan (2011–2015) for energy development, Xinjiang was added to the list. In the meantime, the plan encouraged more renewable power connected to the power grid and restricted the coal power generation in the eastern of China since climate change and air pollution [36]. Along with the expanding of electricity transmission, VWT embodied in electricity transmission also increased significantly from 2005 to 2014. A deep investigation of the evolution of the VWT network is vital. Besides, to distinguish the impacts posed by different policies posted on the evolution of the VWT network, we divided the time span into two periods, i.e., 2005 to 2010, 2010 to 2014.

2.4. Data Sources

This paper analyzed VWT based on electricity transmission among thirty provincial grids in mainland China. The power consumption and generation of Tibet are excluded because Tibet is isolated from other grids. In reality, the Inner Mongolia power grid is divided into the eastern and western parts, which are operated by two companies. Here, to directly investigate the overall transmission of Inner Mongolia, we assemble the eastern and western parts as one provincial grid, i.e., the Inner Mongolia power grid.

In our study, two groups of data are used i.e., electricity data and water consumption of power generation data. We put the data in the Supplementary Materials. As for electricity data, electricity generation data by different power generation technologies at the provincial level are collected from the China Electricity Yearbook, in which electricity consumption data by provinces are also reported. Similar to the previous study [37], pair-wise electricity transmission data are collected from the Annual Complication of Statistics of Power Industry in China [38].

Water consumption for different power generation technologies has been fully investigated at different scales by using different models. Additionally, the same power generation technologies may have different water inventory because of different cooling units. For example, the water consumption of coal-fired power plants differs from different cooling systems, while the water consumption of hydropower is impacted by various factors such as evaporation and season's change. In this study, we collected water inventory of thermal power generation, hydropower generation, and nuclear power generation from [20,22], which fully considered the spatial distribution of different power plants' types. As for wind and solar, we collected the water inventories from previous studies [17,39]. As administrative provinces are the objectives of this study, we estimated the water inventory of provincial electricity generation by:

$$PWC = \frac{\sum\limits_{k=1}^{k=5} EG^k \cdot w_k}{EG} \tag{21}$$

where EG^k is the power generation from k_{th} technology, and w_k is the water coefficient for technologies.

Power generation provinces extracted large amounts of water to satisfy local power plants and virtual water is exported to load hubs when power generation provinces exported its electricity. As the spatial imbalance distribution between energy resources and water resources exists, it is vital to distinguish virtual water delivered from water-abundant provinces and water-scarce provinces. The water stress index (WSI) proposed by [40] is thus used to adjust VW into virtual scarce water (VSW). The WSI indicator could represent the water pressure that a region faced. It is calculated by adjusting the withdrawal-to-availability (WTA) ratio into a constant ranged from 0 to 1, which is shown as follows:

$$WSI = \frac{1}{1 + e^{-6.4 \cdot WTA\left(\frac{1}{0.01} - 1\right)}} \tag{22}$$

The WSI and electricity generation mix of each province is presented in Figure 1, which clearly shows the spatial mismatch between energy resources and water resources. According to the previous study [21], we divide provinces with different WSI into four levels, i.e., no water stress (WSI under 0.2), moderate water stress (WSI 0.2–0.6), serious water stress (WSI 0.6–0.8), and extreme water stress (WIS 0.8–1.0). The spatial distribution of water resources is extremely unbalanced (see Figure 1), with most southern provinces being classed as humid and northern provinces classed as arid. In addition, the spatial distribution of different primary energy resources determines the power generation mix in each province. Even the speed of decarbonization in China's power sector is increasing, the power generated by thermal is still dominated generation mix in most provinces. Most of hydropower plants located in the southern provinces, while other renewable energies located in northwestern and northeastern provinces.

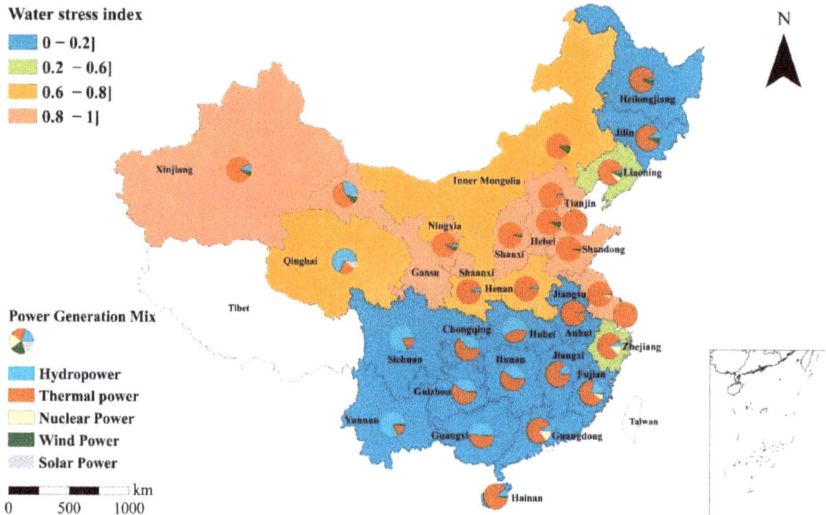

Figure 1. The water stress index (WSI) and electricity generation mix of each province in 2014.

3. Results

3.1. Virtual Water Transfers Embodied in Electricity Transmission

The divergence between power generation provinces and power consumption provinces are increasing, which can be represented directly by the rapidly increasing of VWT's magnitude. The total volume of VWT increased from 450 Mm^3 in 2005 to 5010 Mm^3 in 2014, a factor of 11.3. As for VSWT, it increased from 204 Mm^3 in 2005 to 899 Mm^3 in 2014, a factor of 4.4. The different increasing speed between VWT and VSWT can be ascribed to the imbalance distribution of water resources in China. Additionally, both volumes of VWT and VSWT increased significantly than electricity transmission in the decade, which can be attributed to the expanding of hydropower that owned intensive water consumption.

Similar to the direction of electricity transmission flow, the embodied VWT mainly follows a west-to-east pattern. However, the difference between VWT and VSWT is also obvious because of the different water stress levels and power generation mix between northwestern provinces and southwestern provinces. We use the Circos tool [41] to represent the VWT network in this paper. Figure 2 shows the structure of pair-wise inter-provincial VWT in 2005 and 2014. The five largest VW flows in 2005 are Shanxi-to-Hebei (35 Mm^3), Hubei-to-Jiangsu (34 Mm^3), Hubei-to-Guangdong (30 Mm^3), Jilin-to-Liaoning (30 Mm^3), and Shanxi-to-Beijing (27 Mm^3), while those flow are Yunnan-to-Guangdong

(916 Mm^3), Sichuan-to-Jiangsu (473 Mm^3), Sichuan-to-Shanghai (418 Mm^3), Guizhou-to-Guangdong (384 Mm^3), and Sichuan-to-Zhejiang (336 Mm^3), respectively in 2014. Those top five pair-wise flows' summation account for 51% of all volumetric VWT. In comparison with the VWT between 2005 and 2014, southwestern provinces are playing a more and more important role in the VWT network, especially in Sichuan and Yunnan. However, the structure of VWT in the two provinces is different. In 2014, Sichuan mainly exports VW to the Yangtze River Delta, while Yunnan exports VW to Guangdong. It is ascribed to the distribution of transmission lines constructed between east to the west.

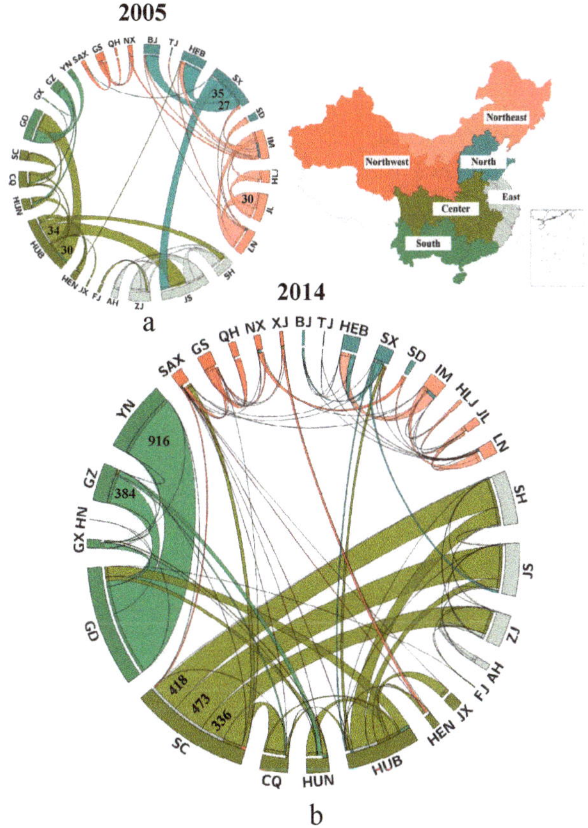

Figure 2. Pair-wise virtual water flows between provinces in 2005 (**a**), and 2014 (**b**). The top five flows are labeled. The color of the ribbon represents the different subnational grid. The ribbon touches inner circle refers to the export, while out of touch means an import. AH—Anhui, BJ—Beijing, CQ—Chongqing, FJ—Fujian, GD—Guangdong, GS—Gansu, GX—Guangxi, GZ—Guizhou, HEB—Hebei, HEN—Henan, HLJ—Heilongjiang, HUB—Hubei, HUN—Hunan, IM—Inner Mongolia, JL—Jilin, JS—Jiangsu, JX—Jiangxi, LN—Liaoning, NX—Ningxia, QH—Qinghai, SAX—Shaanxi, SC—Sichuan, SD—Shandong, SH—Shanghai, SX—Shanxi, TJ—Tianjin, XJ—Xinjiang, YN—Yunnan, ZJ—Zhejiang.

In the view of VSWT, its structure and evolution differ from the counterpart. As is shown in Figure 3, Northern provinces dominated the VWT network, while southwestern provinces are weakened. The top five VSWT flows in 2005 are Shanxi-to-Hebei (36 Mm^3), Shanxi-to-Beijing (27 Mm^3), Shanxi-to-Jiangsu (19 Mm^3), Jiangsu-to-Zhejiang (19 Mm^3), and Jiangsu-to-Shanghai (18 Mm^3), while those are Inner Mongolia-to-Hebei (88 Mm^3), Shanxi-to-Hebei (74 Mm^3), Gansu-to-Qinghai (73 Mm^3), Ningxia-to-Shandong (49 Mm^3), and Sichuan-to-Jiangsu (45 Mm^3), respectively, in 2014.

The summation of top flows accounts for over 30% of the total VWST flows in the network. The comparison of VWT and VSWT could bring us more details. Both northwestern and southeastern provinces are main electricity exporters, but the environmental impacts posed by exporting electricity to local water resources are disproportionate because of different water stress and power generation mix in those provinces. Northeastern provinces are dominated by coal power plants, and the water scarcity in those provinces is much serious, leading to a large amount of VW and VSW run away. In contrast, southwestern provinces are water-abundant and most of the electricity is generated by hydropower plants. Thus, even the water coefficient of hydropower is large compared to thermal power, when VW adjusted to VSW, the volumes of VSW exported from southwestern provinces are very small.

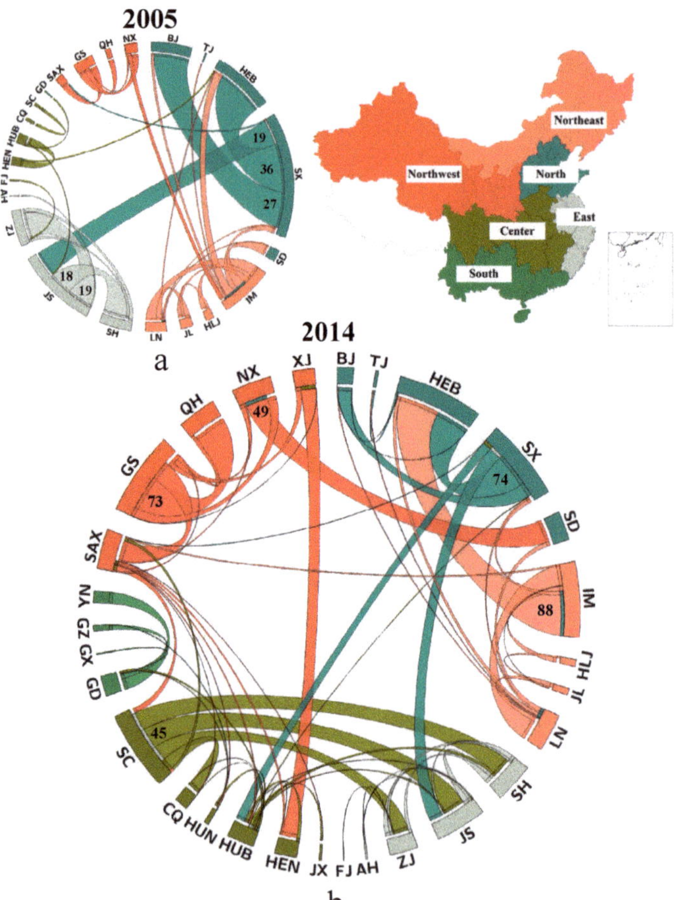

Figure 3. Pair-wise virtual scarce water flows between provinces in 2005 (**a**), and 2014 (**b**). The top five flows are labeled. The color of the ribbon represents the different subnational grid. The ribbon touches inner circle refers to the export, while out of touch means an import. AH—Anhui, BJ—Beijing, CQ—Chongqing, FJ—Fujian, GD—Guangdong, GS—Gansu, GX—Guangxi, GZ—Guizhou, HEB—Hebei, HEN—Henan, HLJ—Heilongjiang, HUB—Hubei, HUN—Hunan, IM—Inner Mongolia, JL—Jilin, JS—Jiangsu, JX—Jiangxi, LN—Liaoning, NX—Ningxia, QH—Qinghai, SAX—Shaanxi, SC—Sichuan, SD—Shandong, SH—Shanghai, SX—Shanxi, TJ—Tianjin, XJ—Xinjiang, YN—Yunnan, ZJ—Zhejiang.

3.2. Driving Forces of Overall Virtual Water Transfers

Different driving factors (i.e., power generation mix, power transmission structure, power generation structure, power demand structure, and power demand) behind the evolution of the VWT network are determined by using a modified SDA model. Decomposition results at the national level for VWT and VSWT are illustrated in Figures 4 and 5 respectively. Overall, changes in the power generation mix, power demand, and power transmission are the main drivers for the increase of VWT in 2005 to 2014. The contribution of the above-mentioned main drivers' change differs between different periods (see Figure 4). For example, the changes in power generation mix contribute to the increase of VWT in 2005 to 2009, and then to the decrease of VWT in 2010 to 2011 and 2012 to 2013. Similar to the power generation mix, the contribution of power transmission changes is different in the decade. Excepted for the period between 2010 and 2011, the change of power transmission contributes to the increase of VWT. The change of power demand always contributes to the increase of VWT because of soaring power consumption in the decade. Furthermore, the contributions from power generation structure and power demand structure were very small compared to the other factors. This phenomenon can be ascribed to two reasons. First, although the electricity transmission increased by 3.1 times from 2005 to 2014, the power generation structure and power demand structure changed a little. In this case, the effects of power generation structure and power demand structure are limited.

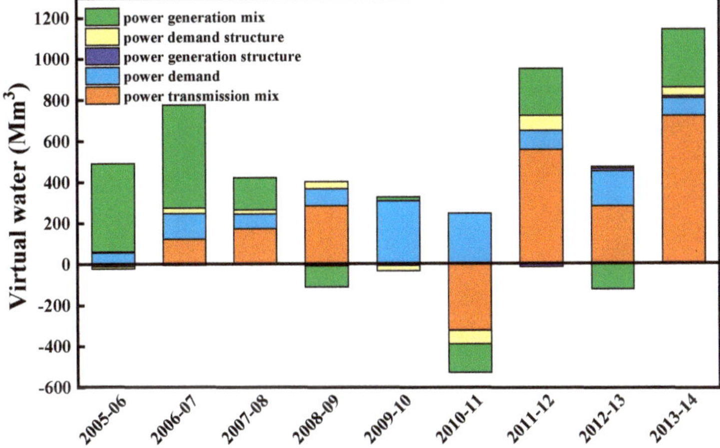

Figure 4. The contribution of driving factors to the changes of virtual water transmission.

In comparison with VWT, the contribution of each factor to the evolution of VSWT is different to some extent. Changes in power demand contribute most to the increase of VSWT while changes in the power generation mix and power transmission also played a vital role in leading the evolution of VSWT. In addition, the extent of effects for factors differs from different periods. For instance, the changes in the power generation mix are the main driver for the decrease of VSWT in 2011 to 2010 (−54%), and 2013 to 2014 (−19%), respectively. The contribution of changes in power transmission is very small (less than 1%) in 2012 to 2013, and it increased to a large share (57%) in 2013–2014. Additionally, changes of power demand structure and power generation structure contribute little to the evolution of VSWT, which is similar to VWT.

To clearly show different policies implemented in different periods, we divide the overall time span (2005–2014) into two periods i.e., 2005 to 2010 and 2010–2014 (see Figure 6). In the first period, the changes in power generation mix contribute most to the increase of VWT (60%) and VSWT (40%), which can be ascribed to the increasing amount of hydropower plants in this period. To reduce the air pollution of the power system, China encourages to improve the share of renewable energies, especially hydropower, as the cost of solar and wind power is expensive at that period. In the second period, the changes of power transmission (51%) are the main driver to the increase of VWT while it is power demand (61%) in VSWT. Additionally, the change of power generation mix contributes to the increase of VWT, but to the decrease of VSWT. It can be attributed to the different WSI of provinces.

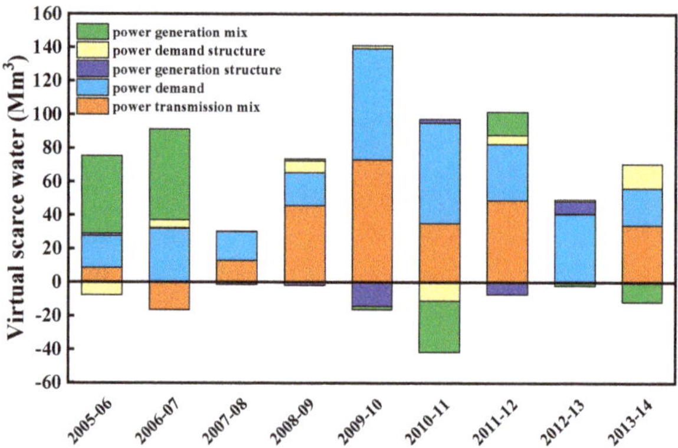

Figure 5. The contribution of driving factors to the changes of virtual scarce water transmission.

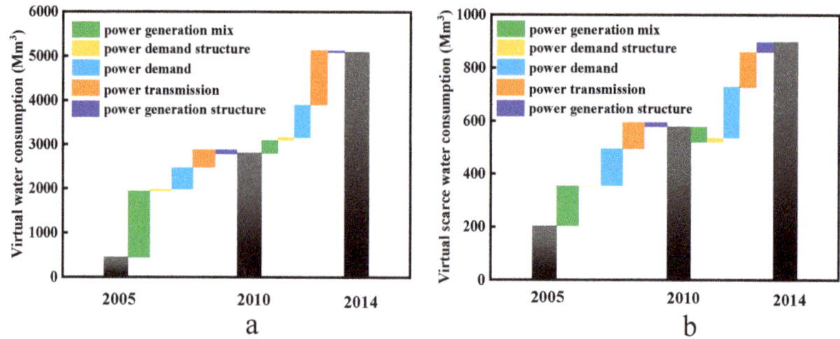

Figure 6. The contribution of different factors to the change of (**a**) virtual water transmission; (**b**) virtual scarce water transmission.

3.3. Driving Forces Analysis at the Provincial Level

The analysis at the national level reveals the overall contribution of factors to the evolution of the virtual water network but covers the details about the contribution of factors to the evolution of the virtual water network in each province. Thus, we investigated the effects of factors on the evolution of VWT and VSWT in the top ten VW and VSW exporters, respectively. The top ten VW exporters in 2014 are Sichuan, Yunnan, Hubei, Guizhou, Inner Mongolia, Shanxi, Gansu, Anhui, Guangxi, and Hunan, in which the VW exporting accounts for 91% of the total VWT. Additionally, at the provincial level, power generation structure and power demand structure should be considered as external power generation structure and external power demand structure based on modified SDA [32].

The factor decomposition results for the top ten VW exporters from 2005 to 2014 are shown in Figure 7. The contributions of factors differ from various provinces and periods. Generally, power demand, power transmission and power generation mix jointly determined the evolution of the VWT network. Sichuan is the largest VW exporter in 2014. In Sichuan, the changes in power generation mix (114%, 104%) dominated the increase of VWT in 2005–2006 and 2006–2007, while the changes of power transmission contribute mostly to the increase of VWT between 2011–2012 (89%) and 2013–2014 (85%). Excepted Sichuan, the same condition occurs in other southwestern exporters such as Hubei,

Yunnan, and Guizhou. In those provinces, the power transmission is the main driver of the increase of VWT (see Figure 7) in some periods, but it can be the main driver of the decrease of VWT in other periods. For example, in the 2009–2010 and 2010–2011 periods, the power transmission contributes mostly to the decrease of VWT in Hubei. Compared to southwestern exporters, northwestern exporters show some different characters. Power transmission dominated the change of VWT, which is similar to the southwestern provinces, but the change of power demand replaced the power generation mix and contributes to the second-largest increase of VWT. Especially, between 2013 to 2014, the change of external power generation structure (36%) contributes to the most increase in Inner Mongolia while the change of power transmission is the main driver (25.9%) of the VWT's decrease.

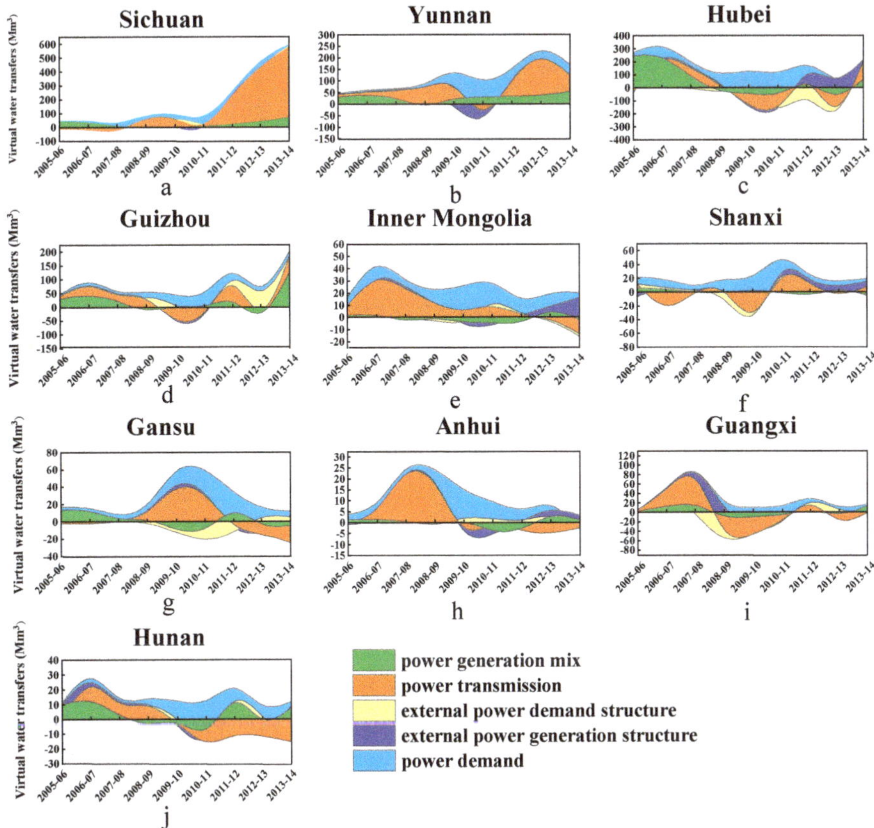

Figure 7. (**a–j**) The contribution of driving factors to the change of virtual water transmission in the top ten virtual water exporters in 2014.

The top ten VSW exporters are Inner Mongolia, Shanxi, Sichuan, Gansu, Ningxia, Xinjiang, Shaanxi, Yunnan, Liaoning, and Hubei, in which the accumulation of VSW accounts for 88.9% of the total VSWT. Figure 8 shows the decomposition results of the VSWT network. The contribution of each factor differs from different periods and provinces. Inner Mongolia is the largest VSW exporter in 2014, and the power transmission contributes mostly to the increase of the VSWT between 2005 to 2009, but it is replaced by power demand in other periods. Similar to the VWT evolution, power generation mix, and power transmission play different roles in different periods. As for the change of external power generation structure and external power demand structure, it generally plays a small role in driving the change of VSWT. However, the change of external power generation structure contributes largely to the change of VSWT in a specific period. For example, it contributes mostly to the increase of VSWT in Inner Mongolia between 2013 and 2014, but to the decrease of VSWT in Ningxia between 2013 and 2014.

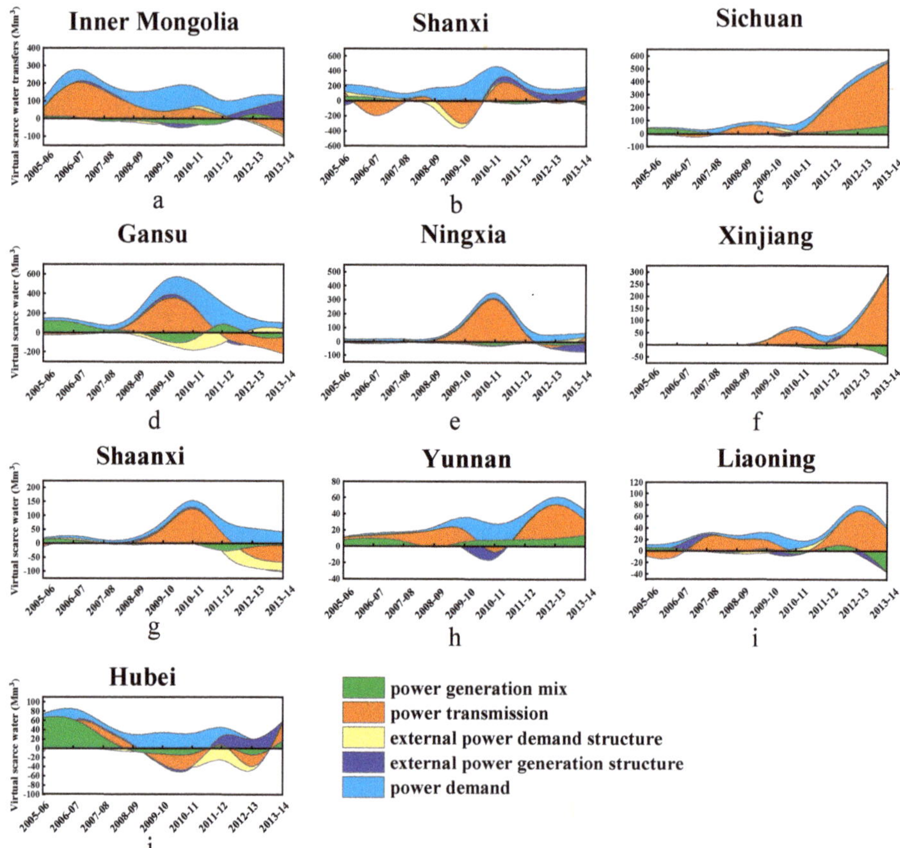

Figure 8. (**a**–**j**) The contribution of driving factors to the evolution of virtual scarce water transmission in the top ten exporters in 2014.

To further investigate the impacts posed by different policies on the evolution of VSWT, we further divided the decade into two periods, i.e., 2005 to 2010 and 2010 to 2014. Figure 9 shows the contributions of factors for the top ten VW and VSW exporters in the two periods. In the first period (here is 2005 to 2010), the power generation mix and power transmission are the main drivers to the change of VWT. In the second period (here is 2010–2014), power demand and power transmission contribute mostly to the change of VWT. However, the change of power transmission in one province (i.e., Shanxi) plays a negative role in the change of VWT in the first period, while it increased to five provinces in the second province. This phenomenon can be ascribed to the evolution of the transmission structure. In comparison with the first period, the power generation structure and the power demand structure play a more important role in the second period, especially in Inner Mongolia and Anhui.

Figure 9 also shows the decomposition results for the top ten VSW exporters in the two periods. In the first decade, Inner Mongolia is the largest VSW exporter, and the change of power transmission (64%) contributes mostly to the increase of VSWT. In the second decade, the power demand replaced the power transmission and plays the largest positive role (75%) in the evolution of VSWT in Inner

Mongolia. The transform between the two periods can be ascribed to the different policies implemented in different periods. Especially, in the first period, the change of power transmission contributes mostly to the decrease of VWT and VSWT in Shanxi. In the second period, the power generation mix is the main driver of the VSWT's decrease in Hubei.

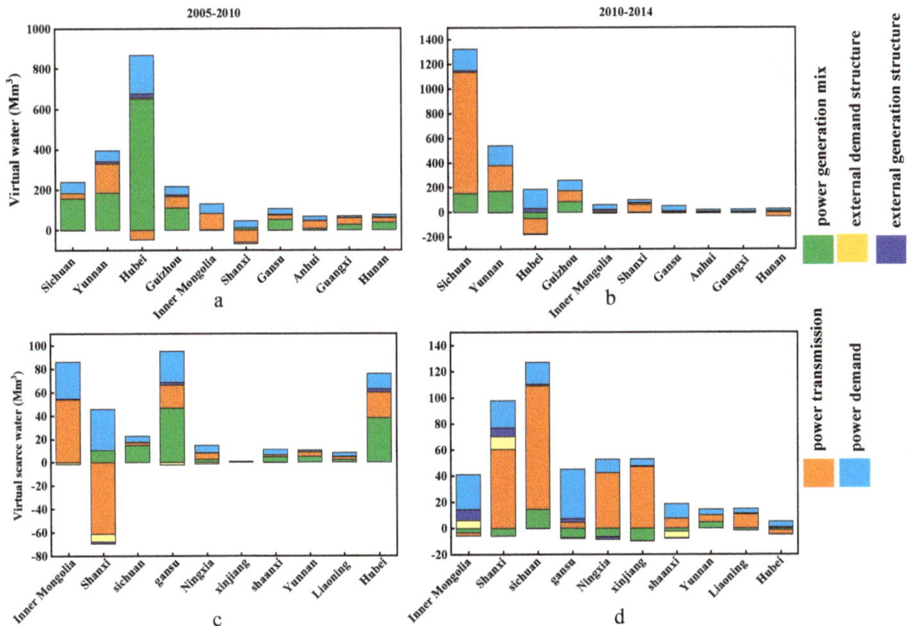

Figure 9. The evolution of virtual water transmission (**a**,**b**), and virtual scarce water transmission (**c**,**d**) of top ten provinces in two periods.

4. Discussion

4.1. Impacts of Policies to the Virtual Water Transmission

The evolution of the virtual water network embodied in China's power system was impacted by different policies implemented over periods. The decomposition model identified the change of power generation mix was a major factor impacting the evolution of VWT from 2005 to 2010 (Figure 9). The high proportion of thermal power generation has induced serious air pollution and made negative effects on climate change. The construction of hydropower plants was promoted to meet the power demand of developed regions. During the first period, the power generation from hydropower had increased from 397 TWh to 722 TWh, with a factor of 1.8. However, the water coefficient of hydropower is higher than that of thermal power, which drives more virtual water delivered from southwestern areas to load hubs concentrated in northeastern areas.

The power transmission and power demand are important factors that increased the virtual water transfers in the second period (2010–2014). They attributed to the increase of investment in power transmission lines. The maturity of UHV technologies exceeded the construction of the UHV-power transmission line. In the period, UHV electricity transmission in China had developed rapidly in relation to the long-distance transmission of alternating current and direct current electricity. In 2010, the 1000-kV Nanyang-Jingmen UHV AC project and the 800 kV UHV DC Yunnan-Guangdong project were put into operation. Besides, the 800 kV UHV DC transmission line from Xiangjiaba to Shanghai commenced operation. The UHV transmission project from Jinping to Southern Jiangsu was put into operation in 2012. The expand of transmission capacity made the power grid more feasible and released the hydropower potential of southeastern provinces, including Yunnan, Sichuan, Hubei, and Guizhou. The change of power transmission dominated the increase of VWT at national level, but that was the main driver of the decrease of VWT in terms of the individual province (e.g., Hubei). Even the overall transmission capacity increased, the amount of power transmission from Hubei significantly decreased in this period. It is noted that the power transmission directly from Sichuan and Guizhou can satisfy the power demand in Guangdong and Yangtze River Delta, which was one reason for the decrease of power export in Hubei. The contributions from the changes in power generation and demand structure were little when compared with the aforementioned factors. It can be referred that power generation and demand structure change had no direct relationship with power transmission change.

4.2. Advice for the Development of China's Power System

Centralized spatial distribution of power generation and investment on the UHV project has led to the diverging between power generation and consumption. Driven by the soaring electricity consumption in load hubs, the magnitudes of electricity transmission will continuously increase in the future. For load hubs, importing electricity from western provinces could mitigate their water stress and air pollution, especially the already polluted Jing-Jin-Jiarea [42]. In the view of power generation provinces, exporting electricity can release their power generation potential and develop the local economy. However, the expanding of electricity transmission increasingly reallocated water resources in both the generation side and consumption side, which may aggravate the water scarcity in China because of the mismatch between water resources and energy resources.

We explored the evolution of VW transfers embodied in provincial electricity transmission and identified the related driving factors. According to the results, a large amount of VW was transferred from western to the eastern provinces, which exacerbated the water stress in western provinces. The water consumption of power plants competed with the water consumption of urbanization and agriculture. Many VW exporters suffering serious hydrological challenges [43]. For instance, Inner Mongolia exported VW to Jing-Ji-Jin area for decades, but Inner Mongolia faced high water stress pressure and the challenge of groundwater depletion. Exporting VW from Inner Mongolia exacerbated the conflicts between different sectors. Some northeastern exporters (i.e., Shanxi, Shaanxi, Ningxia, Xinjiang, and Gansu) facing the same conditions. The power-related VW in southwestern areas was mainly delivered to the Yangtze River Delta and Guangdong. In comparison with northwestern provinces dominated by thermal power, hydropower played a vital role in the southwestern provinces. The transfer of VW in hydroelectricity did not pose water stress to the local ecosystem because of rich abundant water resources in southwestern provinces. The transmission pattern of southeastern provinces is thus more positive than that of northeastern provinces.

Considering the competition for the demands of water resources among energy, urban consumption, and agriculture sectors [44], we suggest that policymakers should integrate across the water-electricity nexus at regional and national levels. To reduce the water pressure in China, the government issued the stringent regulation in 2012, which constrained the national freshwater withdrawals into a definite magnitude (670 billion Mm^3). Although water-saving measures have been implemented on a single plant or sector, synergetic management taking water-electricity nexus into account has not yet been implemented by the government.

Instead of the water scarcity, other environmental impacts (e.g., carbon emission, air pollution, etc.) induced by power systems has raised attention worldwide as well [45]. How to deal with those problems with new technologies (renewable energies connecting technology, artificial intelligence technology, etc.) is associated with the development of power grids. Improving the share of renewable energies is a highlighted way to reduce pollution and emission.

4.3. Advantages and Limitations

This study constructed a virtual water network and explored its evolution, based on the long-time series of electricity transmission data. Compared to previous studies using MRIO tables [30], we directly used the power transmission table, which could avoid the bias from sector aggregation and monetary inhomogeneity in MRIO analysis. The water intensity data for power generation technologies dismissed the impacts of spatial distribution. For example, the special climate is various, and different temperatures and wind speed influence the water evaporation of the reservoir. Moreover, the water coefficient for thermal power plants had been reduced with the improvement of cooling systems in the decade [37].

5. Conclusions

This study investigated the dynamics of the virtual water network embodied in the interprovincial electricity transmission in China and identified its driving factors in different periods by dividing the decade into the two-periods of 2005–2010 and 2010–2014. First, the transfer of virtual water and virtual scarce water generally followed a west-to-east pattern, and their magnitude increased rapidly because of the soaring electricity demand. Considering spatial distribution, virtual water exporters deeply influenced the virtual scarce water network, because of various water stress degree. The export of virtual scarce water in northwestern areas was mainly driven by electricity consumption in Jing-Jin-Ji area, and it was mainly driven by electricity consumption in southeastern areas (e.g., Yangtze River Delta and Guangdong). Besides, the newly constructed transmission lines between Xinjiang and Henan made the Xinjiang-to-Henan virtual scarce water flow increased significantly. According to the water endowments in southeastern provinces and northwestern provinces, the virtual water transmission from southeastern provinces benefits the overall water resources, but that from northwestern provinces would aggravate regional water scarcity. Increasingly developed hydropower generation in the southeastern province can mitigate water scarcity in China even it is water-intensive.

The contribution of driving factors differs from different periods because of the implemental effects of policies. At the national level, the change of power generation mix dominated the increase of the virtual water network in the first period (2005–2010), while the change of power transmission dominated the increase in the second period. At the provincial level, power generation mix and power transmission were the main drivers of the change of virtual water transfers in the top ten provinces. Considering the development trend of the power system in the future, the investigation for co-benefits of dealing with different environmental impacts should be highlighted.

Supplementary Materials: The following are available online at http://www.mdpi.com/1996-1073/13/2/328/s1.

Author Contributions: Conceptualization, Y.Z. and S.H.; Methodology, S.H.; Investigation, S.H. and J.W.; Writing—Original Draft, S.H.; Writing—Review & Editing, J.W., Y.Z. and J.L.; Funding Acquisition, J.L. and H.Z.; Resources, C.Z.; Supervision, J.L. All authors have read and agreed to the published version of the manuscript.

Funding: This work was supported by the National Natural Science Foundation of China (51867003), the Basic Ability Promotion Project for Yong Teachers in Universities of Guangxi (2019KY0046; 2019KY0022), the Natural Science Foundation of Guangxi (2018JJB160056; 2018JJB160064; 2018JJA160176), the Guangxi thousand backbone teachers training program, the boshike award scheme for young innovative talents, and the Guangxi bagui young scholars special funding.

Conflicts of Interest: The authors declare no conflict of interest.

References

1. Fang, D.; Chen, B. Linkage analysis for water-carbon nexus in China. *Appl. Energy* **2018**, *225*, 682–695. [CrossRef]
2. Larsen, M.A.D.; Drews, M. Water use in electricity generation for water-energy nexus analyses: The European case. *Sci. Total Environ.* **2019**, *651*, 2044–2058. [CrossRef] [PubMed]
3. Li, J.S.; Chen, G.Q. Water footprint assessment for service sector: A case study of gaming industry in water scarce Macao. *Ecol. Indic.* **2014**, *47*, 164–170. [CrossRef]
4. China State Council (CSC). Opinions of the State Council on the Implementation of the Strictest Water Resources Management System. 2012. Available online: http://www.gov.cn/zwgk/2012-02/16/content_2067664.htm (accessed on 18 December 2019).

5. Ministry of Water Resources of China. China Water Resources Bulletin in 2018. Available online: http://www.gov.cn/xinwen/2019-07/13/content_5408959.htm (accessed on 18 December 2019).

6. Cai, B.; Zhang, B.; Bi, J.; Zhang, W. Energy's Thirst for Water in China. *Environ. Sci. Technol.* **2014**, *48*, 11760–11768. [CrossRef] [PubMed]

7. Nogueira Vilanova, M.R.; Perrella Balestieri, J.A. Exploring the water-energy nexus in Brazil: The electricity use for water supply. *Energy* **2015**, *85*, 415–432. [CrossRef]

8. Farfan, J.; Breyer, C. Combining Floating Solar Photovoltaic Power Plants and Hydropower Reservoirs: A Virtual Battery of Great Global Potential. *Energy Procedia* **2018**, *155*, 403–411. [CrossRef]

9. Zhang, J.; Lei, X.; Chen, B.; Song, Y. Analysis of blue water footprint of hydropower considering allocation coefficients for multi-purpose reservoirs. *Energy* **2019**, *188*, 116086. [CrossRef]

10. Zaunbrecher, B.S.; Daniels, B.; Roß-Nickoll, M.; Ziefle, M. The social and ecological footprint of renewable power generation plants. Balancing social requirements and ecological impacts in an integrated approach. *Energy Res. Soc. Sci.* **2018**, *45*, 91–106. [CrossRef]

11. Sharifzadeh, M.; Hien, R.K.T.; Shah, N. China's roadmap to low-carbon electricity and water: Disentangling greenhouse gas (GHG) emissions from electricity-water nexus via renewable wind and solar power generation, and carbon capture and storage. *Appl. Energy* **2019**, *235*, 31–42. [CrossRef]

12. He, G.; Zhao, Y.; Jiang, S.; Zhu, Y.; Li, H.; Wang, L. Impact of virtual water transfer among electric sub-grids on China's water sustainable developments in 2016, 2030, and 2050. *J. Clean. Prod.* **2019**, *239*, 118056. [CrossRef]

13. Murrant, D.; Quinn, A.; Chapman, L.; Heaton, C. Water use of the UK thermal electricity generation fleet by 2050: Part 1 identifying the problem. *Energy Policy* **2017**, *108*, 844–858. [CrossRef]

14. Murrant, D.; Quinn, A.; Chapman, L.; Heaton, C. Water use of the UK thermal electricity generation fleet by 2050: Part 2 quantifying the problem. *Energy Policy* **2017**, *108*, 859–874. [CrossRef]

15. Allan, J.A. Virtual Water: A Strategic Resource Global Solutions to Regional Deficits. *Groundwater* **1998**, *36*, 545–546. [CrossRef]

16. Allan, T. (School of O and AS.) Fortunately there are substitutes for water: Otherwise our hydropolitical futures would be impossible. In *Bibliographic Information*; Overseas Development Administration: London, UK, 1993.

17. Feng, K.; Hubacek, K.; Siu, Y.L.; Li, X. The energy and water nexus in Chinese electricity production: A hybrid life cycle analysis. *Renew. Sustain. Energy Rev.* **2014**, *39*, 342–355. [CrossRef]

18. Pfister, S.; Saner, D.; Koehler, A. The environmental relevance of freshwater consumption in global power production. *Int. J. Life Cycle Assess.* **2011**, *16*, 580–591. [CrossRef]

19. Zhang, C.; Anadon, L.D. Life Cycle Water Use of Energy Production and Its Environmental Impacts in China. *Environ. Sci. Technol.* **2013**, *47*, 14459–14467. [CrossRef]

20. Zhu, X.; Guo, R.; Chen, B.; Zhang, J.; Hayat, T.; Alsaedi, A. Embodiment of virtual water of power generation in the electric power system in China. *Appl. Energy* **2015**, *151*, 345–354. [CrossRef]

21. Guo, R.; Zhu, X.; Chen, B.; Yue, Y. Ecological network analysis of the virtual water network within China's electric power system during 2007–2012. *Appl. Energy* **2016**, *168*, 110–121. [CrossRef]

22. Zhang, C.; Zhong, L.; Liang, S.; Sanders, K.T.; Wang, J.; Xu, M. Virtual scarce water embodied in inter-provincial electricity transmission in China. *Appl. Energy* **2017**, *187*, 438–448. [CrossRef]

23. Chini, C.M.; Djehdian, L.A.; Lubega, W.N.; Stillwell, A.S. Virtual water transfers of the US electric grid. *Nat. Energy* **2018**, *3*, 1115–1123. [CrossRef]

24. Zhang, Y.Y.; Fang, J.K.; Wang, S.G.; Yao, H. Energy-water nexus in electricity trade network: A case study of interprovincial electricity trade in China. *Appl. Energy* **2020**, *257*, 113685. [CrossRef]

25. NDRC (National Development and Reform Commission); NEA (National Energy Administration). *Energy Production and Consumption Revolution Strategy*; NDRC and NEA: Beijing, China, 2016. Available online: http://www.ndrc.gov.cn/gzdt/201704/t20170425_845304.html (accessed on 18 December 2019).

26. Yan, D.; Lei, Y.; Li, L. Driving Factor Analysis of Carbon Emissions in China's Power Sector for Low-Carbon Economy. *Math. Probl. Eng.* **2017**, *2017*, 4954217. [CrossRef]

27. Hoekstra, R.; van den Bergh, J.C.J.M. Comparing structural decomposition analysis and index. *Energy Econ.* **2003**, *25*, 39–64. [CrossRef]

28. Goh, T.; Ang, B.W.; Su, B.; Wang, H. Drivers of stagnating global carbon intensity of electricity and the way forward. *Energy Policy* **2018**, *113*, 149–156. [CrossRef]

29. Liao, X.; Hall, J.W. Drivers of water use in China's electric power sector from 2000 to 2015. *Environ. Res. Lett.* **2018**, *13*, 094010. [CrossRef]

30. Zhang, Y.; Chen, Q.; Chen, B.; Liu, J.; Zheng, H.; Yao, H.; Zhang, C. Identifying hotspots of sectors and supply chain paths for electricity conservation in China. *J. Clean. Prod.* **2020**, *251*, 119653. [CrossRef]

31. Cai, B.; Zhang, W.; Hubacek, K.; Feng, K.; Li, Z.; Liu, Y.W.; Liu, Y. Drivers of virtual water flows on regional water scarcity in China. *J. Clean. Prod.* **2019**, *207*, 1112–1122. [CrossRef]

32. Wang, S.; Zhu, X.; Song, D.; Wen, Z.; Chen, B.; Feng, K. Drivers of CO_2 emissions from power generation in China based on modified structural decomposition analysis. *J. Clean. Prod.* **2019**, *220*, 1143–1155. [CrossRef]

33. Zhang, C.; He, G.; Zhang, Q.; Liang, S.; Zipper, S.C.; Guo, R.; Zhao, X.; Zhong, L.; Wang, J. The evolution of virtual water flows in China's electricity transmission network and its driving forces. *J. Clean. Prod.* **2020**, *242*, 118336. [CrossRef]

34. Dietzenbacher, E.; Los, B. Structural Decomposition Techniques: Sense and Sensitivity. *Econ. Syst. Res.* **1998**, *10*, 307–324. [CrossRef]

35. NDRC (National Development and Reform Commission). *The 13th Five-Year (2016–2020) Development Plan for Electric Power Industry*; NDRC: Beijing, China, 2017. Available online: https://en.ndrc.gov.cn/ (accessed on 18 December 2019).

36. Liu, M.; Huang, Y.; Ma, Z.; Jin, Z.; Liu, X.; Wang, H.; Liu, Y.; Wang, J.; Jantunen, M.; Bi, J.; et al. Spatial and temporal trends in the mortality burden of air pollution in China: 2004–2012. *Environ. Int.* **2017**, *98*, 75–81. [CrossRef] [PubMed]

37. Zhang, C.; Zhong, L.; Wang, J. Decoupling between water use and thermoelectric power generation growth in China. *Nat. Energy* **2018**, *3*, 792–799. [CrossRef]

38. CEC (China Electricity Council). *Annual Compilation of Statistics of Power Industry*; China Electricity Council: Beijing, China, 2005–2014. (In Chinese)

39. Li, X.; Feng, K.; Siu, Y.L.; Hubacek, K. Energy-water nexus of wind power in China: The balancing act between CO_2 emissions and water consumption. *Energy Policy* **2012**, *45*, 440–448. [CrossRef]

40. Pfister, S.; Koehler, A.; Hellweg, S. Assessing the Environmental Impacts of Freshwater Consumption in LCA. *Environ. Sci. Technol.* **2009**, *43*, 4098–4104. [CrossRef] [PubMed]

41. Krzywinski, M.; Schein, J.; Birol, İ.; Connors, J.; Gascoyne, R.; Horsman, D.; Jones, S.J.; Marra, M.A. Circos: An information aesthetic for comparative genomics. *Genome Res.* **2009**, *19*, 1639–1645. [CrossRef]

42. Wang, Y.; Li, M.; Wang, L.; Wang, H.; Zeng, M.; Zeng, B.; Qiu, F.; Sun, C. Can remotely delivered electricity really alleviate smog? An assessment of China's use of ultra-high voltage transmission for air pollution prevention and control. *J. Clean. Prod.* **2020**, *242*, 118430. [CrossRef]

43. Liao, X.; Chai, L.; Xu, X.; Lu, Q.; Ji, J. Grey water footprint and interprovincial virtual grey water transfers for China's final electricity demands. *J. Clean. Prod.* **2019**, *227*, 111–118. [CrossRef]

44. Chen, B.; Han, M.Y.; Peng, K.; Zhou, S.L.; Shao, L.; Wu, X.F.; Wei, W.D.; Liu, S.Y.; Li, Z.; Li, J.S.; et al. Global land-water nexus: Agricultural land and freshwater use embodied in worldwide supply chains. *Sci. Total Environ.* **2018**, *613–614*, 931–943. [CrossRef]

45. Guo, Y.; Chen, B.; Li, J.; Yang, Q.; Wu, Z.; Tang, X. The evolution of China's provincial shared producer and consumer responsibilities for energy-related mercury emissions. *J. Clean. Prod.* **2019**, *245*, 118678. [CrossRef]

Article

Membrane Capacitive Deionization for Cooling Water Intake Reduction in Thermal Power Plants: Lab to Pilot Scale Evaluation

Wim De Schepper [1,*], **Christophe Vanschepdael** [2], **Han Huynh** [2] and **Joost Helsen** [1]

[1]   VITO nv, Boeretang 200, 2400 Mol, Belgium; joost.helsen@vito.be
[2]   ENGIE Lab, Rodestraat 125, 1630 Linkebeek, Belgium; christophe.vanschepdael@engie.com (C.V.); ngochan.huynhthi@engie.com (H.H.)
*   Correspondence: wim.deschepper@vito.be

Received: 13 February 2020; Accepted: 6 March 2020; Published: 11 March 2020

Abstract: Cooling of thermal power stations requires large amounts of surface water and contributes to the increasing pressure on water resources. Water use efficiency of recirculating cooling towers (CT) is often kept low to prevent scaling. Partial desalination of CT feed water with membrane capacitive deionization (MDCI) can improve water quality but also results in additional water loss. A response surface methodology is presented in which optimal process conditions of the MCDI-CT system are determined in view of water use efficiency and cost. Maximal water use efficiency at minimal cost is found for high adsorption current (2.5 A) and short adsorption time (900 s). Estimated cost for MCDI to realize maximal MCDI-CT water use efficiency is relatively high (2.0–3.1 € m^{-3}_{evap}), which limits applicability to plants facing high intake water costs or water uptake limitations. MCDI-CT pilot tests show that water use efficiency strongly depends on CT operational pH. To allow comparison among pilot test runs, simulation software is used to recalculate $CaCO_3$ scaling and acid dosage for equal operational pH. Comparison at equal pH shows that MCDI technology allows a clear reduction of CT water consumption (74%–80%) and acid dosage (63%–80%) at pH 8.5.

Keywords: cooling tower; response surface model; water; power plant

---

1. Introduction

Energy production accounts for 10% of freshwater withdrawal globally [1]. However, a much larger fraction of total freshwater withdrawal is used for energy production in industrialized countries, e.g., USA (50%), Western Europe (50%) and China (86%) [1,2]. The majority of withdrawal is used for cooling in thermal energy production [3]. The consumption of cooling water in thermal power plants depends highly of geographical location, cooling and fuel type [4] and is generally high for nuclear, coal and gas fired plants. The use of large amounts of fresh water for power production contributes to the increasing pressure on local water resources [2–5]. Reducing the freshwater withdrawal for cooling is expected to result in a substantial reduction in the water footprint of the energy sector. In general, four types of cooling systems can be employed for electricity generation including once-trough, recirculating, dry and hybrid cooling [5]. Dry cooling relies on air as the coolant medium eliminating water withdrawal and consumption totally. The capital cost of dry cooling is approximately ten times higher than that of once-through cooling [5]; as a result, its use is generally restricted to cases where insufficient make up water supply is available. Once-trough cooling using fresh water is less favored due to its large thermal emission to the surface water body; and [6] the number of power plants utilizing wet (evaporative) cooling systems with an open recirculating cooling tower has therefore rapidly increased [7]. Recirculating cooling towers operate in a feed and bleed mode. Circulating water is evaporated to reject heat while fresh feed water is continuously added, and a fraction of the circulating water is discharged as blowdown. Water withdrawal required for cooling depends on the maximum salt concentration that can be maintained in the recirculation water, which is typically limited by operational aspects such as mineral precipitation and scaling [5]. Scaling involves the precipitation of partially water-soluble salts such as calcium carbonate ($CaCO_3$), which is driven by increasing calcium concentration, pH or alkalinity due to evaporation. The $CaCO_3$ hardness of surface water used for cooling depends highly on the geohydrology of the aquifer and can range from soft (<60 mg/L $CaCO_3$) to very hard (>180 mg/L $CaCO_3$). Previously explored strategies to improve water efficiency of wet cooling towers include feed pretreatment [8,9], use of alternative feed sources [10], circulation water conditioning and acidification [11] and blowdown recuperation [12–14].

Feed pretreatment can be achieved with several processes or process trains including conventional clarification, ion exchange and membrane filtration. Membrane based water treatment technologies have essential advantages over ion exchange in terms of environmental indicators but produce more effluent [8]. Membrane capacitive deionization (MCDI) is an emerging electromembrane process that makes use of electrostatic adsorption to remove ions from a feed stream. An MCDI cell consists of two carbon electrodes covered with ion exchange membranes and separated by a flow channel. During purification, a feed stream is applied to the flow channel, and an electric potential is applied to the electrodes. Ions of opposite charge are attracted and electrosorbed in electrical double layers in the anode and cathode, respectively, and thus removed from the feed stream. Accumulation of ionic charge on the electrode increasingly compensates the applied potential until electrode regeneration is required. During regeneration, the electric polarity is reversed to cause desorption of adsorbed ions. This cycle of purification and regeneration produces two streams, desalinated water and brine. Ion exchange membranes are placed in front of the electrodes in MCDI to prevent that during regeneration ions of opposite charge are attracted from the bulk fluid. This would result in incomplete regeneration leading to a reduced electrode adsorption capacity and longer regeneration times during the next purification step [15]. Application of MCDI for surface or brackish water desalination is characterized by low energy consumption, high water recovery and low fouling propensity [16] in comparison to pressure driven processes. A recent pilot study from Tan et al. [17], using a similar MCDI module as used in this work, showed the possibility of further reducing energy consumption by 30% to 40% using an innovative energy recovery system. MCDI is expected to be less prone to fouling and scaling than other membrane-based desalination technologies [18]. The membranes in MCDI protect the carbon electrodes [19] and due to frequent electrode reversal, build-up of fouling is prevented in a similar way to electrodialysis reversal [15]. MCDI is therefore a potentially highly interesting technology for cooling tower feed pretreatment. MCDI is currently not yet widely applied on a large scale but is considered a viable alternative for partial demineralization of low salinity streams [15,20,21]. A limited number of MCDI pilot studies has been published [15]. Dorji and coworkers performed a pilot scale test with MCDI as alternative for 2nd stage RO in seawater desalination. The results showed that MCDI can effectively remove bromide and dissolved salt at lower energy consumption (0.15 kWh m^{-3}) compared with second stage RO (0.35 kWh m^{-3}) at high water recovery [22]. Van Limpt and Van der Wal [9] performed an MCDI pilot study in which MCDI is used to desalinate tap water as feed for an industrial cooling tower (500 kW) and a residential cooling unit (4500 kW). Chemical savings of up to 85% and water savings up to 28% at low energy consumption (0.11–0.23 kWh m^{-3} produced water) were achieved in this study. MCDI water recovery was limited to 80% to prevent calcium carbonate precipitation. However, a preferential uptake of chloride and calcium (20%) was found resulting in a lower risk of $CaCO_3$ scaling. The authors concluded that the energy consumption was similar to what is expected from reverse osmosis (RO). A similar range of energy consumption for RO is mentioned by Qin et al. [23] who developed a mathematical model to compare the energetic performance of MCDI and brackish water RO (BWRO). They concluded BWRO to be significantly more energy efficient than MCDI, at high salt rejections and moderate to high water salinities.

The potential of MCDI for the reduction of cooling water intake in thermal power plants is studied in this paper. An experiments-based approach is used to evaluate the combined performance of

MCDI and a cooling tower. Lab scale experimental data is used in a response surface methodology to determine the optimal working conditions of the coupled MCDI-CT system in view of water use efficiency and cost. The resulting optima are reevaluated for real CT feed water samples in MCDI lab tests and subsequently in an MCDI-CT pilot case study.

2. Materials and Methods

2.1. CT Water Efficiency

Recirculating cooling towers operate in a feed and bleed mode. Circulating water is evaporated to reject heat while fresh feed water is continuously added, and a fraction of the circulating water is discharged as blowdown. The water regime of a cooling tower is conventionally expressed in terms of cycles of concentration (COC). *COC* measures the degree to which the solid impurities in the makeup water are concentrated in the recirculating water of an evaporative system due to evaporation of water. *COC* is defined (Equation (1)) as the ratio between chloride concentration (chemical *COC*) in the circulation water and make-up water or in terms of flowrate (physical *COC*) of make-up (Q_{makeup}) and evaporate (Q_{evap}).

$$COC = \frac{[Cl^-]_{circulation}}{[Cl^-]_{makeup}} \cong \frac{Q_{makeup}}{Q_{makeup} - Q_{evap}} = \frac{1}{1 - Q_{evap}/Q_{makeup}} \tag{1}$$

A more conventional parameter for a system's water use is water use efficiency or utilization (U), defined as the ratio of effectively used (evaporated) water flow over feed water flow (Equation (2)).

$$U = \frac{Q_{evap}}{Q_{in}} \tag{2}$$

COC is the reciprocal of $(1 - U)$ and therefore a non-linear and less appropriate measure of CT water consumption (Equation (1)). For the coupled MCDI-CT system (Figure 1) it is therefore preferred to use U as a measure for water consumption (Equation (2)).

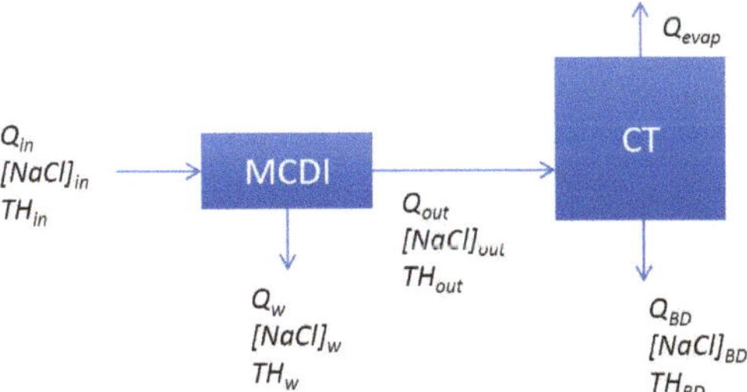

Figure 1. Process scheme for membrane capacitive deionization (MCDI) pretreatment of cooling tower including mass balance components flow (*Q*), [NaCl] and hardness (*TH*) for MCDI feed (*in*), product (*out*) and waste (*w*) and cooling tower evaporate (*evap*) and blowdown (*BD*).

For the MCDI-CT system, U is found as the ratio (Equation (2)) of evaporate flow and MCDI intake water flow (Q_{in}). The maximal achievable utilization (U_{max}), i.e., the maximal fraction of intake water flow that can be used for evaporation, is of specific interest when comparing the efficiency of MCDI-CT under various settings and feed water types. U_{max} is limited by both MCDI water recovery and the maximal CaCO$_3$ concentration that can be achieved in the cooling tower not causing scaling (Equation (3)). For practical use, this can also be expressed in terms of MCDI process characteristics.

$$U_{max} = \frac{Q_{evap,max}}{Q_{in}} = WR\left(1 - \frac{TH_{out}}{TH_{BD,\,max}}\right) \tag{3}$$

where U_{max} is determined by maximal achievable evaporation flow ($Q_{evap,max}$), MCDI intake flow (Q_{in}), MCDI water recovery (WR), hardness of the MCDI treated water (TH_{out}) and maximal allowable hardness in the CT ($TH_{BD,max}$) limited by the maximal solubility of $CaCO_3$. Scaling (by $CaCO_3$) is applied here as single limiting concentration factor for CTs. Several alternative limiting conditions for CT water usage can obviously be envisaged including discharge limitations, material technical limitations (e.g., corrosion) and non $CaCO_3$ related scaling/fouling. These additional limitations could be implemented however following a similar approach.

2.2. MCDI Lab Tests: Setup

The lab-scale setup consists of an Enpar Inc. (Guelph, ON, Canada) lab-scale MCDI cell (0.7 m^2 electrode surface) coupled to a Voltea MCDI circulation loop (Sassenheim, the Netherlands). The system is equipped with a DC power supply, pump, conductivity probe, valves, flow meter and a laptop with LabView-based control and data logging software. During operation, the MCDI system passes through cycles of purification and regeneration. During the purification step, charge builds up in the electrodes as ions are adsorbed from the water. Regeneration is achieved by reversing the polarity over the MCDI cell. All experiments are carried out under fixed flow rate, adsorption current and adsorption time. At the start of each experiment, the MCDI-module is shorted and rinsed with feed water for at least 30 min (at adsorption flowrate) until outlet conductivity equals feed conductivity. In each experiment at least 10 cycles (adsorption-desorption) are completed in flow-through mode (no recycle). Desorption is performed at constant voltage (-1.2 V) or at maximal desorption current (110 A) if -1.2 V is not reached; desorption flow rate is equal to adsorption flow rate and constant; maximal current is applied during desorption; the required desorption time is derived from the charge balance. Feed water is filtered (Pall profile II filter cartridge, 5 µm) prior to testing. During experiments pH, EC (µS cm^{-1}), ΔV (V), Q (mL min^{-1}) and I (A) are continuously monitored. Samples are taken from the influent, purified and waste streams during the last 3 cycles and analyzed for Cl (discrete analysis system and spectrophotometric detection), Na and Ca (ICP-OES).

2.3. MCDI Lab Tests: Experimental Design

MCDI parameter screening is performed on synthetic cooling water following a design of experiments (DOE) approach. Synthetic cooling water is prepared from demineralized water, pro analysis grade NaCl (Merck, Darmstadt, Germany) and $CaCl_2$ ($\geq$94% Merck). The chosen experimental design [24] is a half fractional central composite design (CCD) with 5 factors at 2 levels, i.e., a $2^{(5-1)}$ design, and star points ($\alpha = 1.719$, 30 runs, 3 center points). This type of design consists of a fractional factorial design with center points and is augmented with a group of 'star points' to allow estimation of curvature. The star points are at distance alpha (α) from the center and represent extremes for the low and high settings for all factors. The design is of resolution V, indicating no main effect or two-factor interaction is aliased with any other main effect or two-factor interaction, but two-factor interactions are aliased with three-factor interactions. α is computed for orthogonality using Dell Inc. (2015) Statistica software (data analysis software system, version 12). The design contains 5 factors of which three are MCDI operational factors (adsorption phase time (t_{ads}), adsorption current (I_{ads}), adsorption phase flowrate (Q_{ads})) and 2 factors are related to feed water composition ($[NaCl]_{in}$, feed water hardness (TH_{in})). Factor ranges are selected based on previous experience and real cooling tower feed water qualities (Table 1).

Primary (directly measured) response variables are product water composition ($[NaCl]_{out}$, TH_{out}) and MCDI water recovery (WR). Secondary (calculated) response variables are specific energy use (E, kWh m$^{-3}_{in}$), estimated cost ($Cost$, € m$^{-3}_{in}$) both expressed relative to the feed water flow and selectivity (S, -). Cost estimation is based on the experimental data and standard cost data (unit cell cost: 150 € m^{-2} electrode, E-cost: 0.1 € kWh^{-1}, cell lifetime: 2 years, installation depreciation: 10 year at

4% rate of investment, yearly maintenance cost: 5% of investment). Where required electrode surface (A_{el}) is determined from lab cell electrode surface (A_{cel}) and flow (Equation (4)).

$$A_{el} = \frac{A_{cel}}{Q_{in}} \tag{4}$$

Selectivity [25] is calculated as the ratio of the molar ratio of hardness of the MCDI feed water and the MCDI product water (Equation (5)).

$$S = \frac{\dfrac{M_{CaCO3}\ TH_{in}}{M_{CaCO3}\ TH_{in} + M_{NaCl}\ [NaCl]_{in}}}{\dfrac{M_{CaCO3}\ TH_{out\ (M)}}{M_{CaCO3}\ TH_{out} + M_{NaCl}\ [NaCl]_{out}}} \tag{5}$$

where M_i (g mol^{-1}) indicates the molar masses of $CaCO_3$ and NaCl, respectively. Least-squares multiple regression analysis (Anova) is applied to determine the functional relationship between factors and responses using following polynomial equation (Equation (6)):

$$Y = b_0 + \sum_i b_i X_i + \sum_{ij} b_{ij} X_i X_j + \varepsilon \tag{6}$$

where Y represents a response variable, b regression coefficients, X factors, ε experimental error and i, j running variables. The significance of the generated response surface (RS) equations and model terms are evaluated from Anova table ($\alpha > 0.05$), adjusted coefficient of performance ($R^2_{adj.}$), residuals distribution and Pareto analysis. Insignificant terms are removed in a stepwise model reduction procedure until a fully reduced model is obtained [26]. Experimental design and statistical analysis are performed with Dell Inc. (2015) Statistica (data analysis software system), visualization with Matlab 2016 version 12 (The MathWorks, Natick, MA, USA) and data preprocessing with MS office (Microsoft).

Table 1. Experimental design: factor levels and ranges.

Factors	Factor Levels and Ranges ($\alpha = 1.719$)				
	$-\alpha$	-1	0	1	α
$[NaCl]_{in}$ (ppm)	10	530	1255	1980	2500
TH_{in} (ppm)	30	150	315	480	600
t_{ads} (s)	250	930	1875	2820	3500
I_{ads} (A)	0.50	1	1.75	2.5	3
Q_{ads} (mL min^{-1})	50	92	150	210	250

2.4. CT Feed Water

Cooling tower feed water samples are collected from 3 existing thermal power plant locations (Brussels-Charleroi (BC) Canal, Belgium; Gent-Terneuzen (GT) Canal, Belgium; Eume river, Spain) and treated municipal sewage water (STP effluent; Mol, Belgium). MCDI tests are performed with real cooling water samples to compare the effect of different feed water types on MCDI-CT process parameters (E, S, U_{max} and $Cost$). Following process settings are used: (t_{ads}, I_{ads}) = (900 s, 2.5 A) and (t_{ads}, I_{ads}) = (2000 s, 1 A), for each setting two flowrates are selected (60 and 120 mL min^{-1}).

2.5. MCDI-CT Pilot

The MCDI-CT pilot installation consists of a 30 m^3 feed tank, MCDI container and a mobile cooling tower unit (Merades). The MCDI 10-foot container holds a prefiltration device (5 µm bag filter, Filtermat) coupled to a Voltea CAP-DI IS2 MCDI unit with a capacity of 0.2–1.8 m^3 h^{-1}. The pilot is fully automated (Siemens PLC) and remotely controllable, Selenium webdriver with Python is used for data collection. 0.1 M NaOH and 0.1 M citric acid solutions are used for cleaning in place (CIP). Merades consists of two parallel and independent cooling tower loops simulating semi-open cooling circuits. Each circuit consists of a cooling tower basin, circulation pump and a 12 m length condenser pipe. Water is pumped from the cooling water basin through the condenser where water temperature is slowly increased, condenser outlet temperature is controlled, and is subsequently sent to the cooling tower where it is cooled before flowing down back into the cooling tower basin. The cooling tower is equipped with a nozzle to spray the water through a fill. The temperature in the cooling tower basin is regulated by varying the forced air flow (ventilator) in the cooling tower. The *COC* of the cooling circuit is regulated by a blowdown pump. Merades is equipped with automated injection system and continuous inline monitoring. Make-up and circulation water are monitored for pH and *EC* (continuously), hardness (TAC and THCa) and chlorides (hourly), LPR corrosion measurement (3 times/week), free and total ATP and total viable count (weekly).

The pilot installation was operated for 6 test runs during a 12-week test period (May–July 2018, Linkebeek Belgium). Feed water is abstracted from Brussels-Charleroi Canal and transported to the pilot location (Engie Laborelec, Linkebeek, Belgium) on ~weekly basis. In each test, Merades loop 1 is fed with untreated Canal water (reference) while loop 2 is supplied with MCDI treated water. *COC* of both Merades loops is gradually increased until scaling occurs to determine the maximal achievable *COC*. Condenser inlet pH is maintained constant at pH 8.0 (H_2SO_4) in each test. Scaling onset is determined from circulating water $[Ca^{2+}]/[Cl^-]$ ratio. When minimal blowdown flowrate is reached (4×10^{-3} m^3 h^{-1}, *COC* 4 to 5) without scaling occurring, pH is gradually increased to induce scaling. Fixed operational CT parameters include condenser outlet T (37 °C), ΔT in condenser (10 °C), CT water T (27 °C), water spread in condenser pipes 1.46 m s^{-1}, circulation flow rate 1.9 m^3 h^{-1}, packing spraying flow rate 8 m^3 m^{-2} h^{-1}, hydraulic halftime 0.25 h. Stainless-steel condenser pipe and film fill (height 1.5 m) are used. Following each test, the cooling water circuits are cleaned (concentrated HNO_3, pH < 2.0 for 2 h minimum). MCDI CIP is performed following each test and intermittently when required. After cleaning the system and modules are thoroughly rinsed with demineralized water.

3. Results and Discussion

3.1. MCDI Response Surface

Measured response variables and calculated response variables are determined from MCDI lab tests (Table 2).

The primary response data include *[NaCl]$_{out}$*, *TH$_{out}$* and *WR*. Comparison of *[NaCl]$_{out}$* and *[NaCl]$_{in}$* shows that feed water NaCl concentration is effectively reduced during MCDI tests. A maximal reduction of 91% in [NaCl] is found (run 20) while median removal equals only 16%. This indicates that overall removal of [NaCl] is relatively low, which is expected when aiming for partial desalination. A similar trend is found for *TH*, maximal removal amounts to 88% while median removal equals 23%. Water recovery is generally high for MCDI tests (67% min. to 85% med. to 95% max.) This is desired as MCDI *WR* is expected to largely affect overall water use efficiency when MCDI is used in combination with a CT. Secondary response variables include specific energy use (kWh m^{-3}$_{in}$), estimated cost (€ m^{-3}$_{in}$) and selectivity (-). The median specific energy use of the MCDI system amounts to 0.18 kWh m^{-3}, which is well within the expected range [9,20]. Maximal *E* (0.58 kWh m^{-3}) is found for test run 20 in which 90% reduction in [NaCl] was obtained. Cost estimates range from 0.59 € m^{-3}$_{in}$ minimum over 0.98 € m^{-3}$_{in}$ median to 2.95 € m^{-3}$_{in}$ maximum. Cost estimates in literature are scarce and range from low, 0.11 \$ m^{-3} for CDI (no membranes) on low salinity feed of ≤2000 ppm assuming a 15-year module depreciation [27], to high, 11.7 € ton^{-1} for a 3 kg m^{-3} [Na$^+$] biomass hydrolysate [28]. It needs to be mentioned that the purpose of cost estimation is to distinguish between the economics of different process settings rather than to mirror the exact cost of the MCDI process. Selectivity is calculated for the different runs and *S* is found to vary between 0.8 minimum and 1.45 maximum with 1.08 as median. Following data acquisition, least-squares multiple regression analysis and a subsequent model reduction procedure are applied resulting in a set of regression equations relating response variables to relevant factors and combinations thereof (Table 3).

Table 2. Central composite design data. Factors (left) and response variables (right) are given for each test run. (C) indicates center point.

Run	Factors					Primary Response Variables			Secondary Response Variables		
	$[NaCl]_{in}$ (ppm)	TH_{in} (ppm $CaCO_3$)	t_{ads} (s)	I_{ads} (A)	Q_{ads} (mL min^{-1})	$[NaCl]_{out}$ (ppm)	TH_{out} (ppm $CaCO_3$)	WR (-)	E (kWh m$^{-3}_{in}$)	Cost (Euro m$^{-3}_{in}$)	S (-)
1	530	480	2790	2.5	92	371	313	0.67	0.23	1.60	1.05
2	530	480	811	1	92	354	322	0.91	0.22	1.60	1.00
3	1980	150	811	1	92	1691	119	0.93	0.19	1.59	1.07
4 (C)	1255	315	1800	1.75	150	1037	244	0.91	0.18	0.98	1.06
5	530	150	2790	1	92	328	68	0.86	0.21	1.60	1.31
6	1980	480	2790	1	92	1676	375	0.88	0.19	1.59	1.07
7	1980	150	811	2.5	210	1616	119	0.85	0.21	0.71	1.03
8	1980	480	811	1	210	1870	401	0.92	0.09	0.70	1.11
9	1255	315	3500	1.75	150	1122	255	0.79	0.11	0.98	1.09
10	1255	600	1800	1.75	150	1046	472	0.86	0.17	0.98	1.05
11 (C)	1255	315	1800	1.75	150	1041	228	0.91	0.17	0.98	1.13
12	1255	315	1800	1.75	250	1086	248	0.85	0.11	0.59	1.09
13	1980	480	2790	2.5	210	1829	410	0.71	0.09	0.70	1.07
14 (C)	1255	315	1800	1.75	150	1063	237	0.91	0.17	0.98	1.11
15	1980	150	2790	1	210	1849	130	0.83	0.08	0.70	1.07
16	530	150	2790	2.5	210	472	114	0.74	0.09	0.70	1.15
17	2500	315	1800	1.75	150	2206	270	0.88	0.17	0.98	1.03
18	1980	480	811	2.5	92	1488	268	0.88	0.41	1.62	1.30
19	530	480	2790	1	210	467	388	0.83	0.10	0.70	1.06
20	530	150	811	2.5	92	49	18	0.82	0.58	1.63	0.80
21	1255	315	1800	0.5	150	1166	269	0.94	0.07	0.97	1.08
22	1255	315	1800	3	150	973	221	0.77	0.19	0.99	1.09
23	1980	150	2790	2.5	92	1737	116	0.72	0.21	1.60	1.13
24	10	315	1800	1.75	150	8	106	0.80	0.26	0.99	1.07
25	530	480	811	2.5	210	385	289	0.78	0.24	0.72	1.13
26 (C)	1255	315	1800	1.75	150	1094	236	0.91	0.18	0.99	1.14
27	1255	315	1800	1.75	50	675	112	0.83	0.55	2.95	1.45
28	530	150	811	1	210	441	96	0.85	0.10	0.70	1.26
29	1255	30	1800	1.75	150	1050	18	0.78	0.17	0.98	1.39
30	1255	315	100	1.75	150	1255	315	0.89	0.19	0.99	1.00

Table 3. Reduced regression equations ($\alpha = 0.05$) resulting multivariate regression on MCDI test data. Statistics include R^2, adjusted R^2, Significance of regression (SOR), lack of fit test (LOF) [29] and sequence of standardized effects ($|t|$), units as in Table 2.

Response	Equation	Anova	Sequence of Effects ($|t|$)
$[NaCl]_{out}$	$= -322 - 202 l_{ads} + 7.34 Q_{ads} + 0.92[NaCl]_{in} - 0.25 t_{ads} + 0.077 t_{ads} l_{ads} - 0.019 Q^2_{ads} + (3.99 \times 10^{-5}) t^2_{ads}$	$R^2 = 0.99$, $R^2_{adj.} = 0.98$, SOR: $[F(6, 23) = 293, p < 0.001]$, LOF: $[F(18, 5) = 16.4, p = 0.0029]$	$[NaCl]_{in}$ (52) $> Q_{ads}$ (5.6) $> Q^2_{ads}$ (4.5) $> t_{ads} l_{ads}$ (3.7) $> l_{ads}$ (3.7) $> t^2_{ads}$ (2.7) $> t_{ads}$ (3.8)
TH_{out}	$= -204 - 30.8 l_{ads} + 2.18 Q_{ads} + 0.91 TH_{in} + 0.118[NaCl]_{in} - 0.086 t_{ads} - 0.081 TH_{in} l_{ads} + (1.99 \times 10^{-2}) t_{ads} l_{ads} - (5.74 \times 10^{-3}) Q^2_{ads} - (3.18 \times 10^{-5})[NaCl]^2_{in} + (1.65 \times 10^{-5}) t^2_{ads}$	$R^2 = 0.97$; $R^2_{adj.} = 0.95$, SOR: $[F(10, 19) = 62.2, p < 0.001]$, LOF: $[F(2, 25) = 18.0, p = 0.018]$	TH_{in} (23) $> [NaCl]_{in}$ (5.1) $> Q_{ads}$ (4.8) $> Q^2_{ads}$ (3.2) $> l_{ads}$ (2.8) $> [NaCl]^2_{in}$ (2.7) $> t^2_{ads}$ (2.6) $> t_{ads} l_{ads}$ (2.3) $> TH_{in} l_{ads}$ (1.5) $> t_{ads}$ (1.5)
WR	$= 8.63 \times 10^{-1} + (1.02 \times 10^{-4})[NaCl]_{in} - (3.01 \times 10^{-8})[NaCl]^2_{in} + (5.01 \times 10^{-5}) t_{ads} - (1.35 \times 10^{-8}) t^2_{ads} + (1.18 \times 10^{-8}) TH^2_{in} - (2.63 \times 10^{-2}) l_{ads} - (4.52 \times 10^{-7}) Q^2_{ads} - (2.32 \times 10^{-5}) t_{ads} l_{ads}$	$R^2 = 0.80$, $R^2_{adj.} = 0.73$, SOR: $[F(8, 21) = 10.8, p < 0.001]$, LOF: pure error $= 0$	l_{ads} (8.3) $> t_{ads}$ (6.3) $> [NaCl]_{in}$ (3.1) $> TH^2_{in}$ (3.0) $> t_{ads} l_{ads}$ (2.4) $> [NaCl]^2_{in}$ (2.3) $> Q^2_{ads}$ (2.1) $> t^2_{ads}$ (1.9)
$Cost$	$= 3.90 - 0.03 Q_{ads} + (7.2 \times 10^{-5}) Q^2_{ads}$	$R^2 = 0.96$; $R^2_{adj.} = 0.96$, SOR: $[F(2, 27) = 309, p < 0.001]$, LOF: $[F(2, 25) = 16.4, p < 0.001]$	Q_{ads} (23) $> Q^2_{ads}$ (9.8)

The equations for the primary response variables $[NaCl]_{out}$ and TH_{out} have a good fit ($R^2_{adj.}$, $R^2 = 0.99$ and 0.95), while that for WR is less good ($R^2_{adj.} = 0.83$). Post-hoc testing of residuals shows that the assumption of normality is satisfied (Shapiro–Wilk's (SW) W test: $[NaCl]_{out}$, $W = 0.96$, $p = 0.40$; TH_{out}, $W = 0.98$, $p = 0.76$; WR, $W = 0.95$, $p = 0.17$). The effect of different factors on response variables is quantified by the pareto order of their standardized effects (Table 3). It can be seen that $[NaCl]_{out}$ depends largely on the $[NaCl]_{in}$. Besides this effect, Q_{ads}, I_{ads} and t_{ads} have a smaller but relevant influence on desalination. Low $[NaCl]_{out}$ occurs for intermediate t_{ads}, high I_{ads} and low Q_{ads}. From factor signs, it can be seen that $[NaCl]_{out}$ is lowest at low flow rates. This is in accordance with previous findings [30], ion removal rate increases with increasing flowrate, but the effect of shorter contact time due to increased flow rate is larger. TH_{out} depends largely on TH_{in} and is also lowest at intermediate t_{ads}, high I_{ads} and low Q_{ads}. Trends for TH_{out} are highly similar to those for $[NaCl]_{out}$. In addition, less hardness is removed from brackish water compared to sweet water, since TH_{out} depends also on $[NaCl]_{in}$. Maximal WR is achieved at low I_{ads} and short t_{ads}, which conflicts with the desired parameter settings required for low TH_{out} and $[NaCl]_{out}$. For the secondary response variables, only the reduced equation for *Cost* has a good fit ($R^2_{adj.} = 0.95$). The equation (Table 3) indicates that *Cost* is depending of Q_{ads} and Q^2_{ads} only. Flowrate is directly related to required electrode surface, which confirms the notion that MCDI cost is largely determined by equipment cost [23]. Residuals analysis shows a systematic underprediction of cost at low flowrate and overprediction at high flowrate causing the normality assumption not to be satisfied (SW, $W = 0.77$, $p < 0.001$). A power law of Q and *Cost* is fit and used instead for further evaluation (*Cost* $= 146$, $Q_{ads}^{-0.997}$, $R^2 = 0.99$). The model equation for E has a less good fit ($R^2_{adj.} = 0.82$) and residuals normality was not satisfied (SW, $W = 0.91$, $p = 0.013$). The model equation for S consists of an intercept only. This is indicative of a small but significant selectivity for Ca^{2+} removal ($\bar{S} = 1.11$; t (4.9); $p < 0.001$). The equations for S and E are not further used in the analysis of the combined MCDI-CT system.

3.2. MCDI-CT Process Evaluation

3.2.1. Water Use Efficiency

MCDI is studied as CT pretreatment in order to maximize water use efficiency. The MCDI process settings that result in the largest water use efficiency are therefore of interest and are determined from the MCDI response surface model (RSM). More specifically U_{max} (Equation (3)) is determined from WR and TH_{out} model equations as function of process settings (Q_{ads}, I_{ads} and t_{ads}) and feed water types. Feed water quality is reduced to four distinct feed water types (sweet/soft, sweet/hard, brackish/soft, brackish/hard) to allow to visualize U_{max}. Hardness limit ($THBD_{max}$; Equation (3)) is set to TH 225 ppm $CaCO_3$ for soft water and TH 600 ppm $CaCO_3$ for hard water. This accords to a maximal COC of 1.5 or a U_{max} of 0.33 for a CT without pre-treatment. Values are based on the operational conditions of existing Belgian power plants located at Brussels-Charleroi Canal and Ghent-Terneuzen Canal. Evaluation of the resulting volume plots (Figure 2) reveals that the highest U_{max} can be obtained with MCDI treated sweet/soft water. Optimal U_{max} for sweet/soft water type ($U_{max} = 0.79$) is the highest achievable value of U_{max} in the evaluated parameter space ($t_{ads} = 900$ s; $I_{ads} = 2.5$ A; $Q_{ads} = 90$ mL min^{-1}). Application of MCDI on sweet/soft water results in the lowest TH_{out} resulting in higher achievable water use efficiency when compared to high salinity or high hardness water types.

Lowest U_{max} values are found for hard/brackish water with a lower optimal $U_{max} = 0.52$ realized at low Q_{ads}, high I_{ads} and intermediate t_{ads}. MCDI on intermediate water types results in intermediate optimal U_{max} (0.57 hard/sweet and 0.62 soft/brakish). An overall trend is observed for U_{max} as function of the studied operational parameters and different feed water types. Within a specific feed water type, flow rate is found to have the largest effect (inverse) on U_{max} (colour gradient varies strongest with Q_{ads}). Comparison amongst different feed water types shows that the dependency of U_{max} on flowrate is strongest for sweet water. For a given flowrate, a ridge type optimum is found stretching from high I_{ads} and low t_{ads} to low I_{ads} and intermediate t_{ads}.

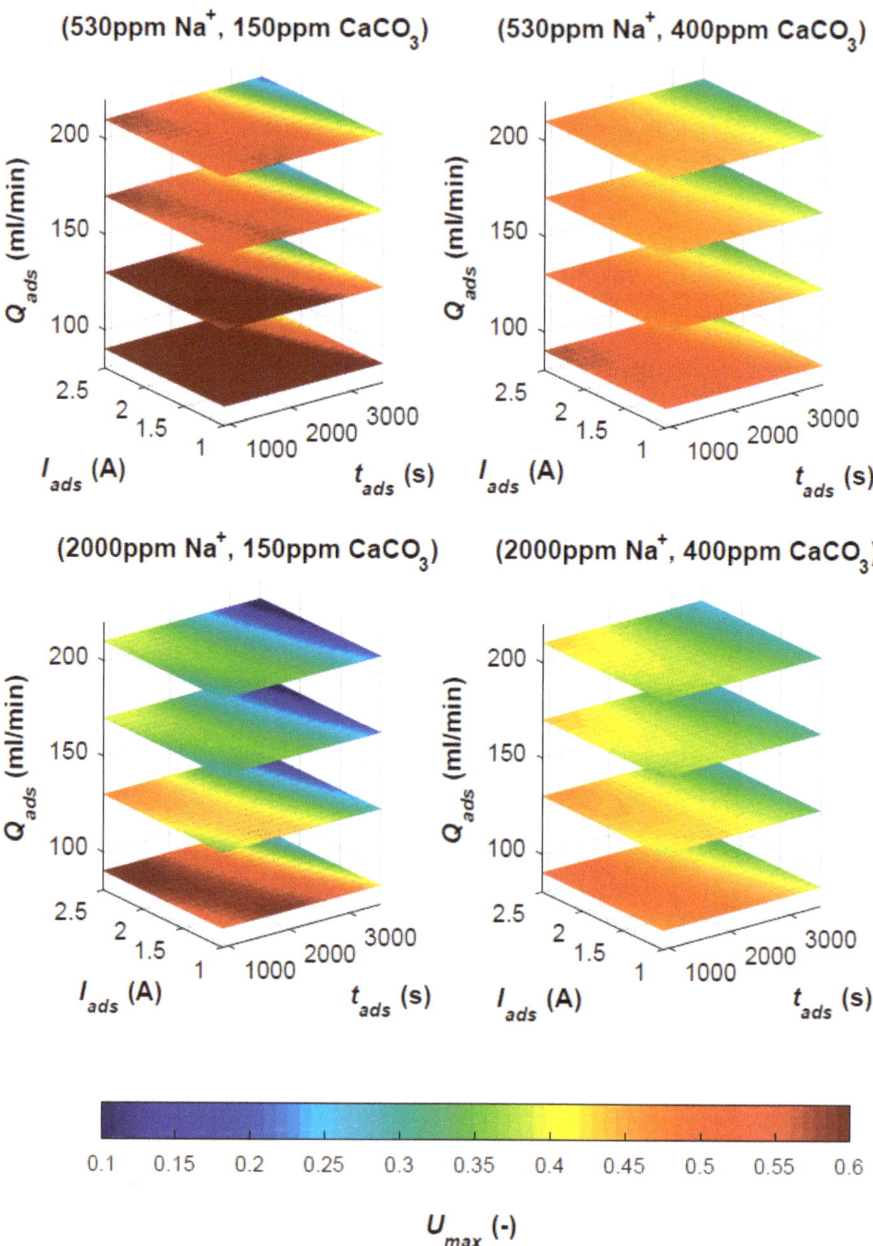

Figure 2. 4D volume plots of maximal utilization (U_{max}) for 4 feed water qualities ([Na$^+$], *TH*). U_{max} is given as colormap slices in the volume plot of flowrate (*Q*), current (*I*) and duration (*t*) for MCDI adsorption phase.

3.2.2. Cost Estimate

Estimated cost is determined from energy consumption and required electrode surface (Equation (4)). Both parameters depend largely on flowrate. Increasing Q_{ads} causes specific energy consumption (*SEC*) and A_{el} to lower causing *Cost* to decrease. With increasing flowrate also U_{max} decreases which makes it of interest to optimize U_{max} and *Cost* in terms of I_{ads} and t_{ads}. MCDI *Cost* is made relative to CT evaporate flow to allow comparison of cooling capacity for the 4 previously defined feed water types. Iso-surfaces are plotted which represent a single optimal U_{max} value as a function of cost (€ m^{-3}evap.), I_{ads} and t_{ads} (Figure 3).

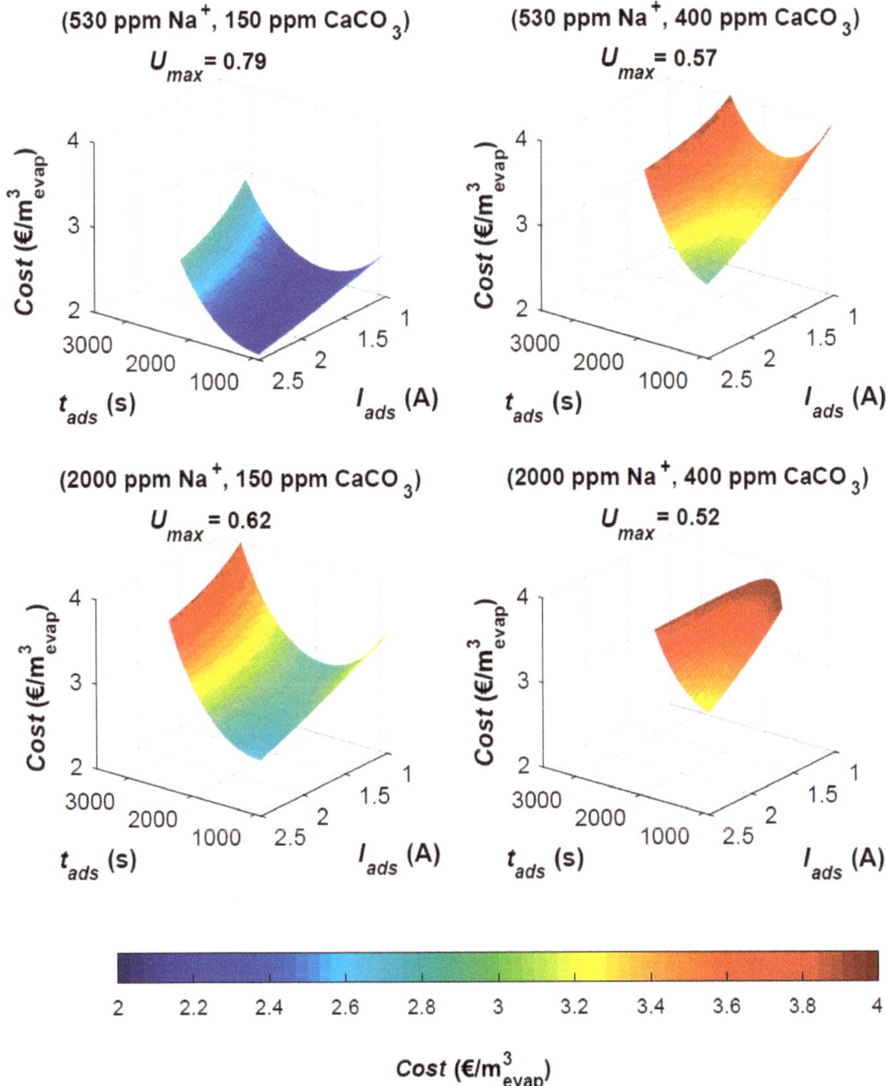

Figure 3. *Cost* (€ m^{-3}evap.) at maximal utilization (U_{max}) versus current (*I*) and duration (*t*) of adsorption phase. For 4 distinct combinations of feed water composition ([Na$^+$], *TH*), colormap indicates cost.

Comparison of different feed water types shows a notable effect of TH_{in} on cost. Cost increases strongly with hardness and to a lesser extent with increasing $[NaCl]_{in}$. For each feed water type, an optimum setting exists for t_{ads} and I_{ads} resulting in lowest cost for a given U_{max}. These optima (Table 4) are found at high I_{ads} combined with low t_{ads} or low I_{ads} combined with intermediate t_{ads} similar to the previously described optima for U_{max}. For sweet/soft feed water, e.g., the optimum extends from low I_{ads} (1 A) and optimal t_{ads} (~2000 s) up to high I_{ads} (2.5 A) and an optimal t_{ads} (<1000 s). The existence of such optimum allows minimizing the cost by choosing optimal I_{ads} and t_{ads} to achieve a given U_{max}. Since cost is determined by Q_{ads}, the objective could also be expressed as maximizing Q_{ads} to reach a specific U_{max}. Q_{ads} is a design parameter that largely depends on installation size for a given application. Alternative optima can be considered, e.g., maximizing U_{max} at fixed cost or minimizing footprint. Moreover, optimization of the product/waste cycle in MCDI can be of use to further maximize water recovery, since WR relates directly to U_{max} (Equation (2)).

Table 4. Optimal U_{max} for membrane capacitive deionization cooling towers (MCDI-CT) system (from screening tests) and corresponding operational conditions (t_{ads}, I_{ads}, Q_{ads}), specific energy consumption (*SEC*), required electrode surface (A_{el}) and *Cost* (€/per m^3_{evap}) for 4 feed water qualities.

	Water Type	Soft/Sweet	Soft/Brackish	Hard/Sweet	Hard/Brackish
Feed quality	TH_{feed} (ppm CaCO$_3$)	150	150	400	400
	$[Na^+]_{feed}$ (ppm)	530	2000	530	2000
Settings	Optimal U_{max} (-)	0.79	0.62	0.57	0.52
	t_{ads} (s)	900	900	900	900
	I_{ads} (A)	2.5	2.5	2.5	2.5
	Q_{ads} (mL min^{-1})	90	90	90	90
Cost	SEC (kWh m$^{-3}_{evap}$)	0.56	0.71	0.77	0.85
	A_{el} (m^2 h m$^{-3}_{evap}$)	164	209	227	249
	$cost$ (€ m$^{-3}_{evap.}$)	2.0	2.6	2.8	3.1

3.3. MCDI on Real Feed Water

MCDI tests are performed on real CT feed water samples (BC Canal, GT Canal, Eumes River and STP effluent). Treated STP effluent is included as a possible alternative source of cooling water [31]. The selected feed water types have a distinct composition (Table 5). BC Canal water has a relatively low $[Na^+]$ and a medium to high TH ($[Ca^{2+}]$ = 125.7 ppm, $[Mg^{2+}]$ = 15 ppm). GT Canal feed water TH is highly similar to BC Canal ($[Ca^{2+}]$ = 114 ppm, $[Mg^{2+}]$ = 43 ppm) while being higher in $[Na^+]$ (300 ppm Na$^+$). Eumes river feed water is very low in TH ($[Ca^{2+}]$ < 0.05 ppm, $[Mg^{2+}]$ < 10 ppm) with a relatively high $[Na^+]$ and high pH compared to the other water types. STP effluent has a low sodium concentration (22.2 ppm) and relative low TH ($[Ca^{2+}]$ = 23.9 ppm, $[Mg^{2+}]$ = 3.3 ppm). In addition, total organic carbon (TOC) is not considered specifically in this study but could contribute to membrane fouling. Indicative TOC values for the cooling water samples (Table 5) are 70 mg C/L (BC Canal), 4.4 mg C/L (Eumes river) and <15 mg C/L (STP effluent).

Table 5. Chemical composition of cooling water samples.

Water Source	BC Canal	GT Canal	Eumes River	STP Effluent
EC (mS/cm)	0.991	2.54	1.096	0.299
pH (-)	8.30	7.7	10.95	7.56
Ca^{2+} (ppm)	125.7	114	<10	23.9
Mg^{2+} (ppm)	15.0	43	<0,5	3.3
Na$^+$ (ppm)	60.1	300	237.5	22.2
Cl$^-$ (ppm)	98.1	527	8.82	31.0
SO$_4^{2-}$ (ppm)	166.3	265	9.65	32.0
NO$_3^-$, NH$_4^+$ (ppm N)	4.6	5.12	0.415	n.a.
PO$_4^{3-}$ (ppm P)	<0.05	<0.05	<0.05	n.a.

MCDI tests with real CT feed water are used to reevaluate the RSM model. Since both Eumes river and STP effluent water compositions are far outside the factor ranges of the RSM (Table 1) they are not used for comparison with model predictions. RSM ranges for TH_{in} and NaCl are based on the average composition of BC and GT Canal water types. Despite of seasonal variation the current samples (Table 5) have a similar composition (*EC*, [Ca^{2+}]) when compared to the ranges used for RSM ([Ca^{2+}]: 56–180 ppm, *EC*: 1.2 mS/cm–4.6 mS/cm), the current BC Canal water sample however has a relative low [Na^+]$_{in}$. For GT Canal and BC Canal feed water types, U_{max}, *WR* and TH_{out} are calculated from test results and from RSM equations (predicted value and 95% confidence interval, Statistica prediction and profiling tool) and compared (Figure 4).

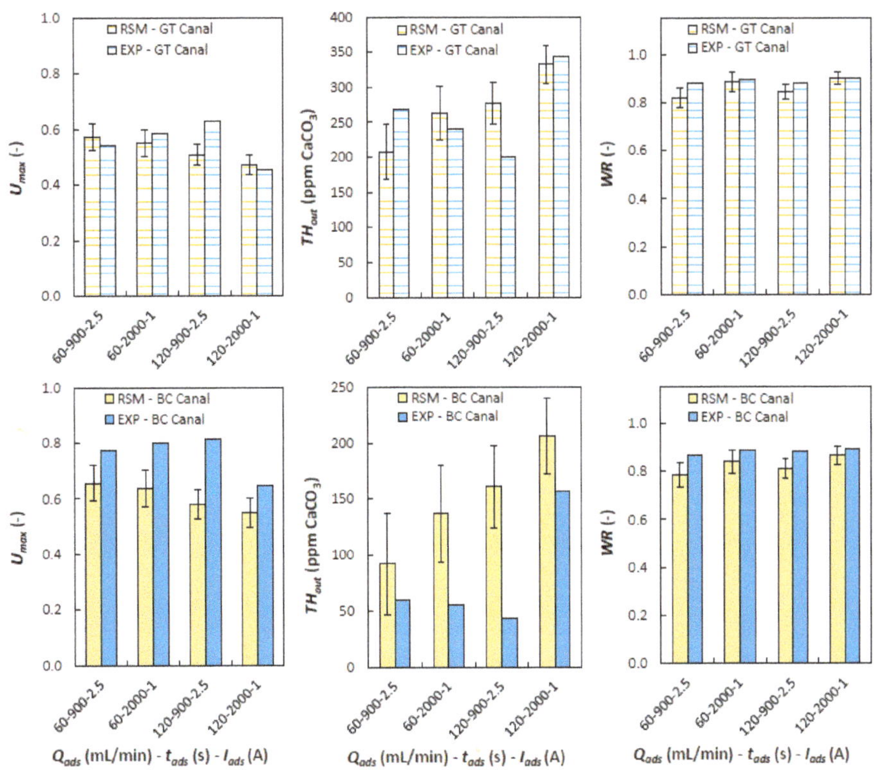

Figure 4. Comparison of U_{max}, *WR* and TH_{out} derived from RSM (±95% CI) and experimental values for GT Canal (**Top**) and GT Canal (**Bottom**) feed water.

Comparison results for GT Canal water shows that U_{max} and *WR* are well predicted by RSM. Prediction of parameter TH_{out} is less accurate and the inverse effect on U_{max} is notable (e.g., for 120-900-2.5). Average difference between RSM predicted and test results derived parameters is <5% for GT Canal water. Comparison of results for BC Canal shows a larger deviation; specifically, TH_{out} is overestimated by RSM. On average, RSM underestimates U_{max} by 20%, *WR* by 6% and overestimates TH_{out} by 89% for BC Canal water. Overestimation of TH_{out} is attributed to the relatively low [Na^+]$_{in}$ in BC channel water (60.1 ppm) compared to the RSM factor range (Table 1) making the RSM model less predictive.

MCDI tests with real CT feed water are further used to evaluate the effect of feed water composition on MCDI-CT process parameters (*SEC*, selectivity, U_{max} and *Cost*). Desired properties of the MCDI process in relation to process efficiency are low cost, low energy consumption, high U_{max} and high selectivity (*S*) for bivalent ion removal (e.g., Ca^{2+}, Mg^{2+}). Process parameters are determined for the 4 selected water types and various MCDI process conditions and plotted for comparison (Figure 5).

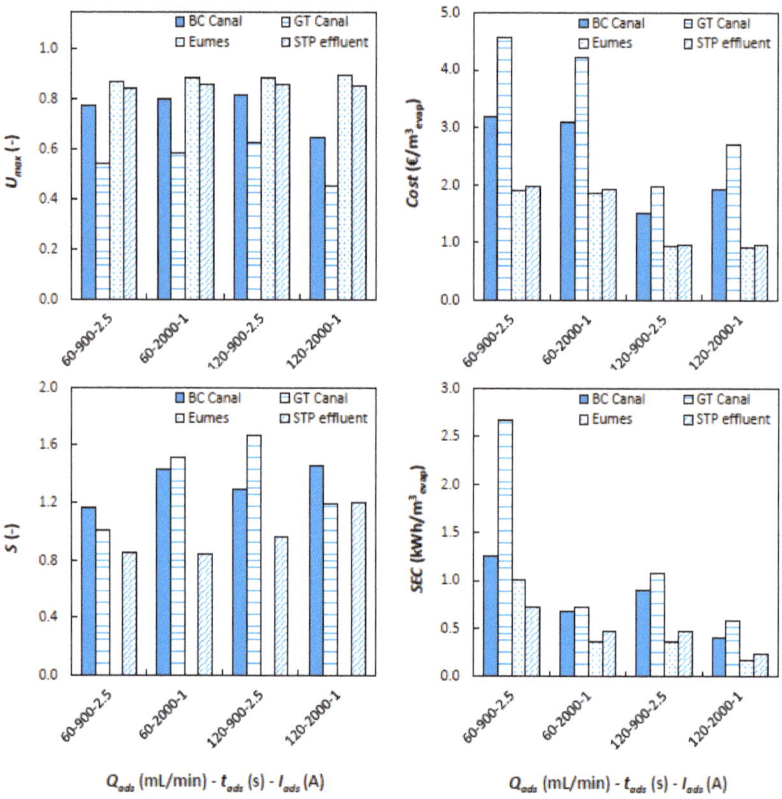

Figure 5. MCDI-CT process parameters *SEC* (kWh/m$^3_{evap}$), U_{max} (-), S (-) and *Cost* (€/per m$^3_{evap}$) derived from MCDI tests on 4 types of real cooling water.

U_{max} is found to be highest for Eumes river water followed by STP effluent, BC Canal and GT Canal feed water types. This expected as U_{max} is inversely correlated with feed water EC and *TH* following their effect on TH_{out}. Variation of U_{max} among process conditions is small, except for BC Canal and GT Canal feed water where ((t_{ads}, I_{ads}):(2000 s, 1 A)) U_{max} is lower in comparison to other test conditions. This is also mirrored in *Cost* (€ m$^{-3}_{evap}$.), which is inversely correlated with flowrate and U_{max} and to a much lesser extent energy consumption (kWh m$^{-3}_{evap}$) as can be seen for BC and GT Canal feed water. Energy consumption (kWh m$^{-3}_{evap}$) is high for BC Canal and GT Canal water types when compared to STP effluent and Eumes feed water. Selective removal of bivalent ions (Ca^{2+} and Mg^{2+}) is a desired MCDI feature. Selectivity is found to some extend for BC Canal water ($\bar{S}$ = 1.35; Standard Deviation (*SD*) = 0.3; N = 4) and GT Canal water ($\bar{S}$ = 1.34; *SD* = 0.13; N = 4). No selectivity was found for STP effluent ($\bar{S}$ = 0.97; *SD* = 0.16; N = 4) while S could not be calculated for Eumes river water due to the lack of hardness ions. Preferential removal of bivalent ions is common in MCDI; it is observed in both synthetic feed mixtures [19,32] and real feed water [9]. This phenomenon is the result of diffusion kinetics and adsorption equilibria in MCDI and is attributed to the preferential storage of multivalent ions in ion exchange membranes [32,33]. Overall comparison shows that high flowrate (120 mL min^{-1}) results in minimal *Cost* (€ m$^{-3}_{evap}$.) for all studied feed water types. Feed water with high TH_{in} (GT Canal, BC Canal) is preferably treated using short t_{ads} while applying high I_{ads}. In addition, the gain in water use efficiency in relation to the base scenario where no pretreatment is in place needs to be considered. For BC Canal and GT Canal water utilization without treatment (U_{max} = 0.33) is relatively low making pretreatment of possible interest while for

STP effluent (U_{max} = 0.76) and Eumes river (U_{max} = 0.99), utilization is already high, and therefore, pretreatment is not useful. Application of MCDI on BC Canal and GT Canal water types results in a relatively high estimated cost per m^3 evaporate. This is partly due to the estimation procedure neglecting in part scale up effects. It generally indicates that the studied MCDI-CT scheme is currently only useful when CT feed water is costly or when legislative boundaries are present that limit water uptake. Legislative constraints on abstraction volumes have been reported to limit energy production in Southern Europe and the US [2]. This is specifically critical in countries where thermoelectric power generation is dominant, and regional water scarcity is a significant concern [1]. BC Canal and GT Canal water feed water cases are currently not severely impacted.

3.4. MCDI-CT Pilot Test

The main purpose of the MCD-CT pilot test is to assess the effect of MCDI treated BC Canal feed water on cooling tower performance and acid consumption. BC Canal water was selected as single feed water source for pilot testing in view of relevance, availability and pilot duration (3 months). Equal ambient conditions for comparison are realized by simultaneously feeding one of Merades cooling tower circuits with untreated BC Canal water (reference) and the other one with MCDI treated BC canal water (Figure 6).

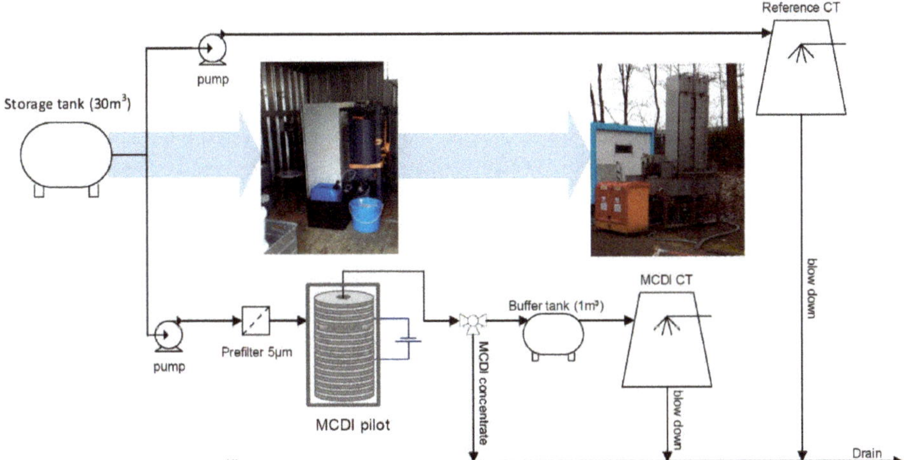

Figure 6. Schematic overview of the MCDI-CT pilot installation. MCDI container with inside view (**left**) and Merades cooling towers and peripheral equipment (**right**).

To allow comparison of both cooling tower circuits, a fixed MCDI product water quality (conductivity setpoint) is produced in each test run by allowing variable current (0–110 A) and fixed flow rate during adsorption phases, yielding a water recovery of 73%–88%. The process was controlled by the setpoint of the conductivity, i.e., either at 25% or 50% reduction of the incoming conductivity. This build in control strategy uses a variable current and adsorption time depending on the product water conductivity. The operating conditions of the pilot therefore differed from the optimal conditions determined from lab tests (Figure 5). In the pilot, specific flowrate was high (10 L/m^{2}h), adsorption time was intermediate (1700 s) and time averaged current density was relatively low (1 A/m^2) for 50% desalination compared to lab tests where following ranges for specific flowrate (5–10 L/hm^2), adsorption time (900–2000 s) and current density (1.5 A/m^2 to 3.5 A/m^2) were used. In practice, removal ratios differed quite significantly in the first 3 test runs due to software issues (Table 6). BC Channel feed water quality is seasonal and both conductivity (0.70–0.87 mS/cm) and [Ca^{2+}] (88–108 ppm) are lower compared to previous samples (Table 5). The specific energy consumption was low (0.08–0.12 kWh m^{-3} produced) in part due to the relative high capacity of the MCDI system used (design flowrate = 0.2–1.8 m^3 h^{-1}, applied flowrate = 0.42 m^3 h^{-1}) but comparable to values found in literature [9,22]. The feed water was also found to have a significant fouling potential previously undetected during lab tests. A steady and consistent increase in hydraulic impedance (10^8 Pa s m^{-3}) of the MCDI cell resulted in an increase in pressure drop over the cell by 2.8 bar in 24 h (at Q_{ads} = 0.42 m^3 h^{-1}). Consequently, the MCDI system passes through an automatic cleaning cycle each 4 h. This type of fouling behavior (spacer fouling) is expectedly caused by small particles and colloidal matter (e.g., clay particles) present in the feed water, which adhere to the spacer fabric, indicating the applied pretreatment (5 μm bag filtration) is insufficient for BC canal feed water. This indicates that fouling prevention remains an important aspect of MCDI operation despite the common notion that MCDI is less vulnerable to fouling compared to other membrane processes as also indicated recently by Choi and coworkers [34]. Overall the MCDI pilot was able to deliver the required volume for each of 6 test runs with Merades CT pilot (Table 6).

Table 6. MCDI removal ratios and operational parameters for Merades cooling tower at test conditions.

Test	Reference	Run 1	Run 2	Run 3	Run 4	Run 5	Run 6
MCDI *EC* removal (%)	0	12	25	43	29	50	50
MCDI *TH* removal (%) *	0	35	19	23	31	52	55
MCDI *WR* (%)	100	82	73	73	88	84	87
CT Operational pH	8.0	8.33	8.2	8.74	8.73	8.7	8.63
CT *COC* (-)	3.95	4.88	4.3	4.75	4.45	4	4.5
U_{max} (-)	0.75	0.65	0.56	0.57	0.68	0.63	0.68

* *TH* removal indicative (point samples).

The resulting U_{max} of the MCD-pilot at operational conditions (Table 6) is lower for MCDI treated water when compared to the reference untreated water. This is deceptive however since a different operational pH is applied in all test runs (no acid dosing). CaCO$_3$ scaling is highly pH dependent, and comparison of CT test conditions therefore requires taking into account both pH control (acid dosage) and water use efficiency explicitly. An extrapolation from test data of CaCO$_3$ scaling, acid dosage and operational pH is performed using ENGIE lab proprietary cooling water simulation software (Table 7).

Table 7. MCDI-CT pilot simulation of COC_{max}, U_{max}, acid dosage (g h^{-1}) and feed water saved versus reference test at same pH for reference and test runs.

Test	Reference	Run1	Run 2	Run 3	Run 4	Run 5	Run 6
pH 8 COC_{max} (-)	3.95	12.9	6.7	31	45	27	35
pH 8 U_{max} (-)	0.75	0.65	0.56	0.57	0.68	0.63	0.68
pH 8 acid dose (g h^{-1})	5.93	4.75	4.05	3.05	3.73	2.64	2.56
Feed water saved (%)	0	−15	−33	−30	−10	−19	−10
pH 8.2 COC_{max} (-)	2.5	6.7	4.3	24	33	20	19
pH 8.2 U_{max} (-)	0.60	0.70	0.56	0.70	0.85	0.79	0.82
pH 8.2 acid dose (g h^{-1})	6.39	4.71	4.03	3.04	3.78	2.57	2.6
Feed water saved (%)	0	14	−6	14	30	24	27
pH 8.5 COC_{max} (-)	1.2	2.85	2.6	9	15	8	7
pH 8.5 U_{max} (-)	0.16	0.53	0.45	0.65	0.82	0.73	0.75
pH 8.5 acid dose (g h^{-1})	10.08	4.43	3.28	2.92	3.75	2.02	2.36
Feed water saved (%)	0	69	63	74	80	77	78

Comparing acid dosage and feed water savings for a given pH shows that MCDI technology allows a clear reduction of water consumption (74%–80%) when CT is operated at higher pH meanwhile strongly reducing acid dosage (63%–80%). MCDI pretreatment reduces acid consumption when operating at low pH but also increases feed water usage by 10%–30%. Under these conditions MCDI pretreatment is less useful. Comparison between operation at pH 8 without MCDI and pH 8.5 with MCDI shows 50% reduction in acid use for comparable U_{max}. It is concluded that the usefulness of MCDI for CT pretreatment depends strongly on operational conditions and feed water type. Specifically, CT feed water with a high hardness benefits from MCDI pretreatment. In addition, monitoring of total viable count (weekly) indicates that MCDI technology has no impact on biological growth in CT recirculation water, suggesting that nutrients (e.g., TOC) required for growth are not extensively removed by MCDI. LPR corrosion measurements show that MCDI treated BC Canal water is more corrosive ($\overline{x}$ = 2.0 mills/year, N = 25) in comparison to non-treated water ($\overline{x}$ = 1.4 mills year^{-1}, N = 25). However, the ability to operate at higher elevated COC and pH counteracts this effect.

4. Summary and Conclusions

The maximal water use efficiency of recirculating wet cooling towers in thermal power production typically depends on feed water composition. Application of MCDI on CT feed water for desalination to increase CT cycles and improve water use efficiency is evaluated in this paper. The combined MCDI-CT process is studied using lab test data following a response surface methodology and mass balancing. Impacts on cooling tower performance and acid consumption are evaluated on pilot scale. The following main conclusions are drawn:

- Response surface modelling shows that feed water type and operational conditions have a major impact on MCDI product water quality. Maximal water use efficiency, U_{max}, depends strongly on MCDI flowrate (Q_{ads}), which is a design parameter. For a given flowrate, optimal U_{max} at minimal cost is found at high I_{ads} (2.5 A) and short t_{ads} (900 s), which relates to process optimization.
- Water use efficiency improves most following MCDI treatment for CT feed water types with high hardness and low initial U_{max} (BC Canal, GT Canal). The effect of MCDI on U_{max} depends strongly on water type.
- Estimated cost for MCDI to realize maximal MCDI-CT water use efficiency is relatively high (2.0–3.1 € m^{-3}$_{evap}$), MCDI is therefore expected to be currently useful only for plants facing high intake water costs or when water abstraction is legally limited. This is expected to become increasingly relevant for water scarce regions that depend on thermoelectric power generation.
- Pilot testing shows that the effect of MCDI pretreatment on water use efficiency depends strongly on CT operational conditions (pH) and feed water type. Water use efficiency is highly pH dependent ($CaCO_3$) and comparison among CT test conditions therefore requires taking into account both pH control (acid dosage) and water use efficiency explicitly using simulation software. This shows that for a CT operating at pH 8.5 on BC Canal water, MCDI pretreatment yields a strong reduction in overall water abstraction (74%–80%) and acid consumption (63%–80%).
- MCDI pretreatment has no impact on biological growth in CT recirculation loop, but treated water is found to be more corrosive. The ability to operate at higher elevated COC and pH is expected to counteract this effect.

Author Contributions: Conceptualization, W.D.S., C.V. and J.H.; data curation, J.H.; formal analysis, W.D.S. and C.V.; investigation, C.V. and J.H.; methodology, W.D.S. and C.V.; project administration, W.D.S. and H.H.; writing—original draft, W.D.S.; writing—review and editing, H.H. and J.H. All authors have read and agreed to the published version of the manuscript.

Funding: This research was funded by The Horizon 2020 program grant number 686031 through the project "Materials and Technologies for Performance Improvement of Cooling Systems in Power Plants (acronym MATChING)".

Acknowledgments: Diane Van Houtven and Ben Jacobs are gratefully acknowledged for their help and discussions.

References

1. Pan, S.; Snyder, S.W.; Packman, A.I.; Lin, Y.J.; Chiang, P. Cooling water use in thermoelectric power generation and its associated challenges for addressing water-energy nexus. *Water Energy Nexus* **2018**, *1*, 26–41. [CrossRef]
2. Byers, E.A.; Hall, J.W.; Amezaga, J.M. Electricity generation and cooling water use: UK pathways to 2050. *Glob. Environ. Chang.* **2014**, *25*, 16–30. [CrossRef]

3.   Larsen, M.A.D.; Drews, M. Water use in electricity generation for water-energy nexus analyses: The European case. *Sci. Total. Environ.* **2019**, *651*, 2044–2058. [CrossRef]

4.   Mekonnen, M.M.; Gerbens-Leenes, P.W.; Hoekstra, A.Y. The consumptive water footprint of electricity and heat: A global assessment. *Environ. Sci. Water Res. Technol.* **2015**, *1*, 285–297. [CrossRef]

5.   Fang, J.; Wang, S.; Zhang, Y.; Chen, B. The electricity-water nexus in Chinese electric trade system. *Energy Procedia* **2018**, *152*, 247–252. [CrossRef]

6.   Raptis, C.E.; Pfister, S. Global freshwater thermal emissions from steam-electric power plants with once-through cooling systems. *Energy* **2016**, *97*, 46–57. [CrossRef]

7.   Peer, R.A.M.; Sanders, K.T. The water consequences of a transitioning US power sector. *Appl. Energy* **2018**, *210*, 613–622. [CrossRef]

8.   Larin, B.M.; Bushuev, E.N.; Larin, A.B.; Karpychev, E.A.; Zhadan, A.V. Improvement of Water Treatment at Thermal Power Plants. *Therm. Eng.* **2015**, *62*, 286–292. [CrossRef]

9.   Van Limpt, B.; van der Wal, A. Water and chemical savings in cooling towers by using membrane capacitive deionization. *Desalination* **2014**, *342*, 148–155. [CrossRef]

10.   Wijesinghe, B.; Kaye, R.B.; Fell, C.J.D. Reuse of treated sewage effluent for cooling water make up: A feasibility study and a pilot plant study. *Water Sci. Technol.* **1996**, *33*, 363–369. [CrossRef]

11.   Altman, S.J.; Jensen, R.P.; Cappelle, M.A.; Sanchez, A.L.; Everett, R.L.; Anderson, H.L.; McGrath, L.K. Membrane treatment of side-stream cooling tower water for reduction of water usage. *Desalination* **2012**, *285*, 177–183. [CrossRef]

12.   Koeman-Stein, N.E.; Creusen, R.J.M.; Zijlstra, M.; Groot, C.K.; van den Broek, W.B.P. Membrane distillation of industrial cooling tower blowdown water. *Water Resour. Ind.* **2016**, *14*, 11–17. [CrossRef]

13.   Frick, J.M.; Féris, L.A.; Tessaro, I.C. Evaluation of pretreatments for a blowdown stream to feed a filtration system with discarded reverse osmosis membranes. *Desalination* **2014**, *341*, 126–134. [CrossRef]

14.   Farahani, M.H.D.A.; Borghei, S.M.; Vatanpour, V. Recovery of cooling tower blowdown water for reuse: The investigation of different types of pretreatment prior nanofiltration and reverse osmosis. *J. Water Process Eng.* **2016**, *10*, 2214–7144. [CrossRef]

15.   AlMarzooqi, F.A.; al Ghaferi, A.A.; Saadat, I.; Hilal, N. Application of Capacitive Deionisation in water desalination: A review. *Desalination* **2014**, *342*, 3–15. [CrossRef]

16.   Pawlowski, S.; Huertas, R.M.; Galinha, C.F.; Crespo, J.G.; Velizarov, S. On operation of reverse electrodialysis (RED) and membrane capacitive deionisation (MCDI) with natural saline streams: A critical review. *Desalination* **2020**, *476*. [CrossRef]

17.   Tan, C.; He, C.; Fletcher, J.; Waite, T.D. Energy recovery in pilot scale membrane CDI treatment of brackish waters. *Water Res.* **2020**, *168*, 115146. [CrossRef]

18.   Anderson, M.A.; Cudero, A.L.; Palma, J. Capacitive Deionization as an Electrochemical Means of Saving Energy and Delivering Clean Water. Comparison to Present Desalination Practices: Will It Compete? *Electrochim. Acta* **2010**, *55*, 3845–3856. [CrossRef]

19.   Hassanvand, A.; Chen, G.Q.; Webley, P.A.; Kentish, S.E. An investigation of the impact of fouling agents in capacitive and membrane capacitive deionisation. *Desalination* **2019**, *457*, 96–102. [CrossRef]

20.   Zhao, R.; Porada, S.; Biesheuvel, P.M.; van der Wal, A. Energy consumption in membrane capacitive deionization for different water recoveries and flow rates, and comparison with reverse osmosis. *Desalination* **2013**, *330*, 35–41. [CrossRef]

21.   Fritz, I.A.; Zisopoulos, F.K.; Verheggen, S.; Schroën, K.; Boom, R.M. Exergy analysis of membrane capacitive deionization (MCDI). *Desalination* **2018**, *444*, 162–168. [CrossRef]

22.   Dorji, P.; Kim, D.I.; Hong, S.; Phuntsho, S.; Shon, H.K. Pilot-scale membrane capacitive deionisation for effective bromide removal and high water recovery in seawater desalination. *Desalination* **2020**, 114309. [CrossRef]

23.   Qin, M.; Deshmukh, A.; Epsztein, R.; Patel, S.K.; Owoseni, O.M.; Walker, W.S.; Elimelech, M. Comparison of energy consumption in desalination by capacitive deionization and reverse osmosis. *Desalination* **2019**, *455*, 100–114. [CrossRef]

24.   Montgomery, D.C. *Design and Analysis of Experiments*, 6th ed.; Wiley & Sons: Hoboken, NJ, USA, 2015; ISBN 0-471-66159-7.

25.   Dong, Q.; Guo, X.; Huang, X.; Liu, L.; Tallon, R.; Taylor, B.; Chen, J. Selective removal of lead ions through capacitive deionization: Role of ion-exchange membrane. *Chem. Eng. J.* **2019**, *361*, 1535–1542. [CrossRef]

26. Cheng, C.D.; Gong, G.W.; Na, L. Response surfacemodeling and optimization of direct contactmembrane distillation for water desalination. *Desalination* **2016**, *394*, 108–122. [CrossRef]

27. Welgemoed, T.J.; Schutte, C.F. Capacitive Deionization TechnologyTM: An alternative desalination solution. *Desalination* **2005**, *183*, 327–340. [CrossRef]

28. Huyskens, C.; Helsen, J.; Groot, W.J.; de Haan, A.B. Cost evaluation of large-scale membrane capacitive deionization for biomass hydrolysate desalination. *Sep. Purif. Technol.* **2015**, *146*, 249–300. [CrossRef]

29. Bezerra, M.A.; Santelli, R.E.; Oliveira, E.P.; Villar, L.S.; Escaleira, L.A. Response surface methodology (RSM) as a tool for optimization in analytical chemistry. *Talanta* **2008**, *76*, 965–977. [CrossRef]

30. Kim, N.; Lee, J.; Hong, S.P.; Lee, C.; Kim, C.; Yoon, J. Performance analysis of the multi-channel membrane capacitive deionization with porous carbon electrode stacks. *Desalination* **2020**, *479*, 114315. [CrossRef]

31. Cherchi, C.; Kesaano, M.; Badruzzaman, M.; Schwab, K.; Jacangelo, J.G. Municipal reclaimed water for multi-purpose applications in the power sector: A review. *J. Environ. Manag.* **2019**, *236*, 561–570. [CrossRef]

32. Hassanvand, A.; Chen, G.Q.; Webley, P.A.; Kentish, S.E. A comparison of multicomponent electrosorption in capacitive deionization and membrane capacitive deionization. *Water Res.* **2018**, *113*, 100–109. [CrossRef] [PubMed]

33. Ahualli, S.; Fernandez, M.; Iglesias, G.; Jimenez, M.L.; Liu, F.; Wagterveld, M.; Delgado, A.V. Effect of solution composition on the energy production by capacitive mixing in membrane-electrode assembly. *J. Phys. Chem. C* **2014**, *118*, 15590–15599. [CrossRef] [PubMed]

34. Choi, O.; Dorji, P.; Shon, H.K.; Hong, S. Applications of capacitive deionization: Desalination, softening, selective removal, and energy efficiency. *Desalination* **2019**, *449*, 118–130. [CrossRef]

Discussion

The Possible Roles of Wastewater Treatment Plants in Sector Coupling

Michael Schäfer [1,*], Oliver Gretzschel [1] and Heidrun Steinmetz [2]

[1]   Institute Water Infrastructure Resources, University of Kaiserslautern, D-67663 Kaiserslautern, Germany; oliver.gretzschel@bauing.uni-kl.de

[2]   Department for Resource Efficient Wastewater Technology, University of Kaiserslautern, D-67663 Kaiserslautern, Germany; heidrun.steinmetz@bauing.uni-kl.de

*   Correspondence: michael.schaefer@bauing.uni-kl.de

Received: 23 March 2020; Accepted: 8 April 2020; Published: 22 April 2020

Abstract: The development of a power system based on high shares of renewable energy sources puts high demands on power grids and the remaining controllable power generation plants, load management and the storage of energy. To reach climate protection goals and a significant reduction of CO_2, surplus energies from fluctuating renewables have to be used to defossilize not only the power production sector but the mobility, heat and industry sectors as well, which is called sector coupling. In this article, the role of wastewater treatment plants by means of sector coupling is pictured, discussed and evaluated. The results show significant synergies—for example, using electrical surplus energy to produce hydrogen and oxygen with an electrolyzer to use them for long-term storage and enhancing purification processes on the wastewater treatment plant (WWTP). Furthermore, biofuels and storable methane gas can be produced or integrate the WWTP into a local heating network. An interconnection in many fields of different research sectors are given and show that a practical utilization is possible and reasonable for WWTPs to contribute with sustainable energy concepts to defossilization.

Keywords: decarbonization; energy concepts; long-term energy storage; power-to-gas; power-to-X

---

1. Introduction

The energy and the water sector are both essential and vital elements of modern life. Water has to be provided and distributed to citizens, as well as discharged and treated with a substantial use of energy. On the other side, water is used to generate power and deliver or recover energy for human purposes. Due to these facts, the challenges are globally to provide these important services. Furthermore, water and energy systems are complex and strongly linked but are mostly operated independently. Facing climate change forces this water-energy nexus into a more environmentally sustainable system for a rapidly growing population on the globe. Therefore, energy efficiency, energy savings and energy recovery have become development goals all over the world. Nevertheless, every country has, on the basis of its power generation composition, its own unique chances, opportunities and obstacles to handle [1–4].

In regard to meeting the targets of the Paris Agreement of 2015 [5] or the United Nation Sustainable Development Goals [4], a system change is needed now, and the world has to act immediately. Although some countries are proactive and take a leading part in promoting renewable energy sources (RES), the actual efforts are, by far, not sufficient. Nevertheless, there are promising developments for the year 2018. The global investments in the new generating power capacity of RES exceeded that in fossil and nuclear power, with a share of 65%. In several countries, the power sector is transforming rapidly, with growth rates of over 10%, and proportions of more than 20% RES in energy production have been reached, like in Uruguay, Germany and the United Kingdom [6]. The gradual extension of RES and the expedited abandonment of fossil and nuclear energy production results in new problems but also new opportunities for power supplies. There is a shift from demand-oriented power generation to a production-driven generation of electrical energy. In an electricity system with a high share of RES, flexibility options are needed to counterbalance fluctuating wind and solar-based power production to maintain the high standards in its supply [7]. For now, there are just a few hours of surplus energy but with proportions of more than 60% RES; times in which supply exceeds demand are increasing significantly. On this basis, there will be a high need of short-term flexibility in the near future to stabilize power grids and further integrate RES into the energy grids [8]. Short-term storage options are classified in the range of seconds up to 24 h (daily storage). These energy surpluses and deficits have to be balanced by flexible energy generators and consumers. Furthermore, long-term storage capacities are needed to provide enough energy in times of deficits on a larger scale. This is caused by longer periods of low amounts of available wind and sun. To compensate these fluctuations and store or generate energy, based on its availability, fundamentally different applications compared to short-term flexibility are required [9].

Energy has always been stored, but the focus and the technologies used have changed. Storage concepts like pumped storage plants, battery or compressed air systems are not suitable for long-term storage—too expensive or cannot provide enough capacities for an extensive use in every country [10]. Additionally, ecological issues and resource scarcity have to be considered. Unlike solar and wind-based energy production, biogas is the only RES that can be directly stored and flexibly adjusted to the production of electrical energy or heat purposes. This is possible due to a flexible power and gas production, which can be realized by biogas plants or wastewater treatment plants (WWTPs) with anaerobic sludge digestion [11–13]. The required storage technologies are available but must be re-evaluated, utilized or combined appropriately to the new challenges and obstacles, especially regarding economic feasibility during the transformation process from a fossil-based energy system to a renewable energy system [9].

Part of a solution in facing climate change could be a comprehensive use of widely spread existing infrastructure, like in the water sector. For short-term flexibility, it has been shown that WWTPs are capable of performing on energy markets without endangering system functionality to stabilize energy grids with its existing aggregates and ensure the further integration of RES into energy grids [14–16]. The capability of providing ancillary services and taking the fluctuation of the electrical energy consumption/generation pattern into account without endangering the WWTPs' systems' functionalities is shown in [16]. Such a contribution applies as well for long-term storage concepts, theoretically shown, e.g., in [15]. In the following, the focus is on practical implementations of long-term storage concepts according to sector coupling, with special focus on its interaction with WWTPs.

Sector coupling describes the interconnection between the sector's heat, gas, mobility, nonenergetic use of fossil resources (e.g., chemistry/industry) and electrical energy under the use of RES due to appropriate technologies [9]. Especially the gas sector provides with its natural gas infrastructure (NGI) a nearly infinite storage option for RES [17,18]. Similar to the NGI, large WWTPs are commonly distributed close to municipal infrastructures with significant synergies in nearly all of the targeted fields of sector coupling, which makes them preferable locations for a transposition [15,19,20].

The available literature is widely scattered across the different addressed topics, and a comprehensive compilation is missing—in particular, with special focus on the water sector and WWTPs. Therefore, the present work provides a systematic review on several fields of sector coupling and their interactions with WWTPs based on the literature. Starting from describing available and reasonable technologies, followed by analyzing the current situation, evaluating the effects of an implementation on the plant and giving examples for actual worldwide, practical applications in WWTPs, the state of the art is summarized. The objective is not only to collect and present knowledge but also provide a basis for decisions to future plant concepts due to evaluating realized projects. This paper refers to synergies, opportunities and downsides that are given to WWTPs as a local energy center in the role of sector coupling and long-term energy storage.

2. Sector Coupling with WWTPs

Sector coupling is realized by the interconnection between RE surplus and the conversion into storable energy forms and is widely called power-to-X (PtX) based on the targeted sector (e.g., power-to-heat, power-to-gas, etc.). There are a lot of Power-to-X projects on the international level, whilst countries with high shares of RES are more endeavoring, simply because they are more affected. Unsurprisingly, countries like Germany take a leading part in development and realized/planned PtX projects [21]. For the German water sector, potential analysis shows that a significant contribution can be provided in terms of ancillary services, as well as gas flow rates, for long-term storage [16,22], which applies for other countries as well.

WWTPs are ideally suited in a special way for handling different energy forms. They are not only flexible (electrical) energy consumers, with their purification aggregates (cf. [16]), but also flexible producers, with their anaerobic sludge digestion and the subsequent needs-oriented production of electrical energy (cf. [12,23]). They are also able to function as a heat sink (e.g., heating of the digestion tanks) and are capable to provide usable heat sources (e.g., waste heat from combined heat and power plant (CHP) units). Furthermore, a wide range of gases can be utilized: Despite a direct use of electrical energy generated by CHP units, the produced digestion gas supplies, with its 35% carbon dioxide (CO_2) content, a valuable and sustainable green CO_2 source for further Power-to-Gas (PtG) applications, like methanization, which enables a direct feed-in of methane (CH_4) into the local gas grid. Moreover, the produced hydrogen (H_2) can be used on the plant as well (e.g., biological methanization) instead of a mere production and feed-in into the natural gas grid. The operation of an electrolyzer for PtG can be integrated in holistic plant solutions considering different gaseous energy forms and heat. The usually unwanted by-product O_2 can also be used for aeration in the biological treatment or further treatment processes such as ozonization for removing micropollutants from the wastewater. This enables synergies and opportunities in both sectors: WWTPs are able to contribute to the energy sector and face their own new challenges in wastewater treatment at the same time (e.g., micropollutant removal).

Figure 1 gives an overview of the multiple utilization paths and the complex interactions with different sectors and the WWTP. This demonstrates that WWTPs are reasonable energy centers, which are located at nearly every urban area. In the following, the focus is on new findings and realized projects in the field of sector coupling explicitly concerning and interacting with WWTPs for the different sectors.

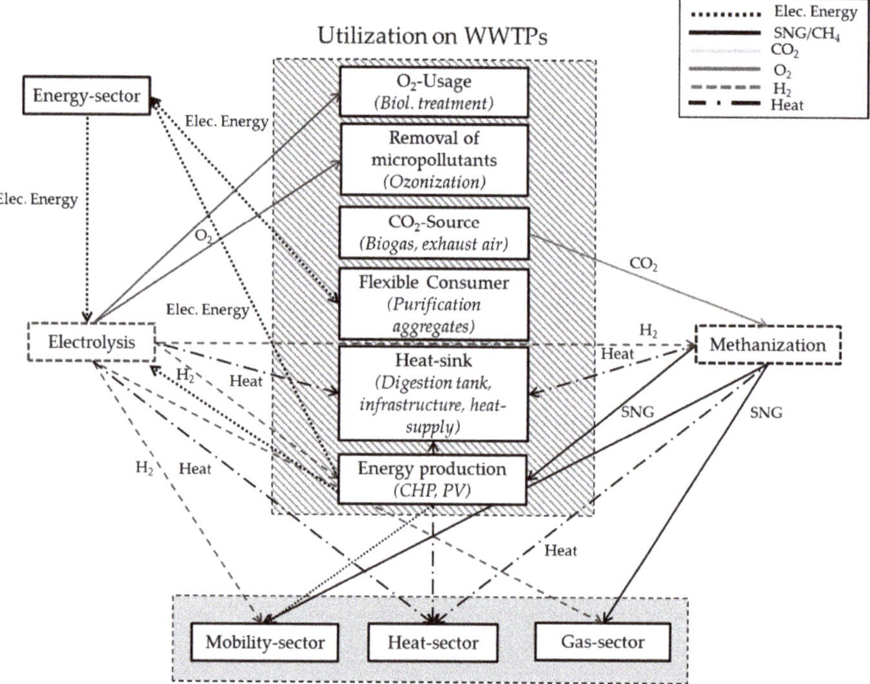

Figure 1. Interaction of wastewater treatment plants (WWTPs) with different technologies and sectors. Reproduced from [16], Technische Universität Kaiserslautern: 2019.

3. Gas-Sector

3.1. Power-to-Gas in WWTPs

A capable long-term storage concept for renewable surplus energy is the transformation into gaseous energy carriers, like H_2 and CH_4, called power-to-gas (PtG). With that technology, energy can be stored and, later on, used again for power production to compensate energetic deficits on a large scale. In a first step, H_2 is produced by the utilization of surplus energy via an electrolyzer, which splits water (H_2O) into H_2 and oxygen (O_2). In a further step, the produced H_2 can be upgraded with CO_2 to CH_4 (Sabatier process). Thus, PtG is divided in two classes: PtG-H_2 and PtG-CH_4. A general comparison of the two concepts is given in Table 1. The core process thereby is the electrolysis, with three different suitable processes: alkaline water electrolysis, polymer-electrolyte membrane electrolysis (PEM) and high-temperature steam electrolysis. Further information and more detailed descriptions on different electrolyzer types can be taken from the report [24]. Whereby the PEM electrolysis seems to be the best-suited option in this framework due to technical advantages regarding fast load changes, good partial-load operations and a good scalability, especially in small-scale implementations (usually below 1 megawatt connected load), like in WWTPs [15,25].

Table 1. General comparison of Power-to-Gas (PtG)-H_2 and PtG-CH_4 concepts (adapted and extended from [9,18,26]).

	Power-to-Gas(-H_2)	Power-to-Gas(-CH_4)
Chemical equation (simplified)	$2H_2O \xrightarrow{\text{Electrolyzis}} 2H_2 + O_2$	$2H_2O \xrightarrow{\text{Electrolyzis}} 2H_2 + O_2$ $4H_2 + CO_2 \xrightarrow{\text{Methanization}(CO_2)} CH_4 + 2H_2O$
Efficiency *	54%–84%	49%–79%
Advantages	• Easier direct use in the mobility sector due to higher market penetration • No CO_2 source needed (more independent in location selection) • Higher efficiency than PtG-CH_4 • Contribution to decarbonization in the industry and mobility sectors due to more customer availability	• Accessible infrastructure available for all fields using natural gas • Easier to compress, to store and to transport due to a triple-higher energy density than H_2 • Nearly infinitely storable due to the NGI • No additional gas treatment needed due to biological methanization
Drawbacks	• Limited feed-in capacity of the NGI (2%–10% volume) • H_2 infrastructure hardly developed	• Additional transformation step needed (higher costs and loss in efficiency rate) • Slightly lower efficiency than PtG-H_2 • Constrains due to available CO_2 sources (e.g., restrictions in location selection, green CO_2 source, etc.)

* Efficiency may vary due to the type of storage, the pressure and the used technology (further information is given in [9]). NGI: natural gas infrastructure.

Nevertheless, the produced gas needs to be stored. The NGI is a huge, easily accessible and well-distributed storage system for long-term storage and providing natural gas to customers. Furthermore, it is one of the best-developed and accessible infrastructures in most countries. A part of the produced H_2 can be directly stored in the NGI (e.g., 2°–10% volume in Germany), and for the CH_4, the storage capacities are nearly infinite due to almost identical chemical attributes of the gases [10,18]. With this background, the surplus energy can be stored, distributed and is accessible for months up to years to compensate seasonal fluctuations in the energy production without building one's own distribution/storage system [9,26].

Therefore, PtG plants have to be located close to the gas grid. As biogas plants are mostly located in rural areas, large WWTPs are present in nearly every urban location, mostly close to the NGI. This makes WWTPs favorable as a PtG location. Another big advantage of WWTPs with anaerobic digestion as a possible location for a PtG plant is the already existing gas infrastructure on-site and the existing know-how on handling gases. On plants with anaerobic sludge digestion, a suitable infrastructure is available in terms of digestion tanks, gas storages and combined heat and power plants (CHP). Furthermore, one of the biggest challenges for PtG(-CH_4) is the lack of an appropriate CO_2 source ("green carbon") [9]. This is a major benefit of WWTPs due to an easily accessible and sustainable biogas with a high CO_2 content of nearly 35% volume [27]. To use the CO_2, it is possible to extract it from the biogas (biogas treatment) or use the raw biogas directly to upgrade the CH_4 content to feed-in quality for the NGI. A techno-economical study of state-of-the-art of such methanization concepts for PtG was conducted by [28] and showed that, especially, biological methanization has a great potential for further development. For WWTPs, a utilization is possible by directly inserting H_2 into the digestion tank or upgrade the biogas to a feed-in quality with a special external reactor [28–31].

Basic plant concepts for PtG implementations in WWTPs and the theoretical feasibility are shown in [15], characteristic values and a feasibility study for upgrading biogas with H_2 are given in [31]. A large-scale practical demonstration for a two-year operation is located at the Avedøre WWTP, Denmark [32] and for a biogas plant in Pirmasens, Germany [30]. Both implementations could demonstrate a successful upscaling from a laboratory scale to a nearly full scale. A continuous feed-in of the produced CH_4 with a high level of flexibility of the biological processes could be established and is running in a stable operation. Nevertheless, an implementation of PtG requires a high control of real-time data and complex coordination of the different parameters, like the CO_2 supply, power supply, operation of the electrolyzer, storage of intermediate gases and utilization of oxygen and heat [32].

Although PtG in WWTPs is technologically possible, much remains to be done. Especially regarding optimization in the coordination of the different material flows and embedding the system in the whole WWTP operation business to gain the full benefits of the synergies. Due to the complexity of process engineering of big WWTPs itself, implementing PtG concepts are quite challenging. However, the greatest problems in implementing are not technology driven but the legal and regulatory framework, which makes the systems so far not feasible without funding (e.g., for Germany: cf. [15,31]). Better conditions would lead to dropping prices, a needed market penetration and faster development for full-scale application.

It can be stated that WWTPs are able to implement PtG systems and can ensure an efficient methanization due to a constant and viable CO_2 source. In addition, it provides a supplement flexibility option for energy grids as a flexible RES. Furthermore, a sustainable and beneficial utilization path for oxygen is given, which leads to unique location advantages for WWTPs.

3.1.1. CH_4 Utilization

Reaching an "energy-positive WWTP" is the proclaimed goal of many operators of anaerobic sludge digestion plants, which can be primary realized by reducing energy consumption, mostly due to efficiency measures, and increasing biogas production. Therefore, the produced digestion gas is mostly used in CHP units in WWTPs to produce electricity. The present literature shows that increasing self-sufficiency is still the dominant topic rather than an interest in holistic and overarching utilization of the different energy forms (e.g., [33–35]). This is mostly caused by missing incentives and the regulatory framework and taxes, which makes selling gas/electricity not economically viable in contrast to a use on-site (e.g., in Germany; [9]). The economic feasibility of anaerobic digestion in WWTPs is highly dependent on the legal framework conditions. A large-scale study of the feasibility of a conversation from aerobic to anaerobic stabilization was conducted in [36]. It could be shown that it is feasible even for WWTPs at sizes of 20,000 PE_{120} under fitting circumstances (for Germany). Furthermore, the boundary conditions favoring conversation were evaluated, but it was also shown that, especially, taxation systems have a huge impact, and feasibility has to be examined individually for each plant and each country's framework [36].

In a future RE-dominated system, biogas should be used to generate electricity with the CHP units if the RE production in the grid is lower than its demand and vice versa. If the demand is below the actual production, the CHP unit would be shut down, and the electrolysis is used to convert the energy surplus into CH_4 and stored in the NGI afterwards. Another possibility of a holistic gas usage could be a more centralized than decentralized electricity production for the CH_4. The (electric) efficiency of modern CHP units in WWTPs is around 43%, with an electrical power scale in the range of 100 up to 1000 kW_{el} [37]. At the same time, big natural gas power plants reach up to 60% [38]. Under a holistic point of view, the overall benefit from the utilization of the gas would be much higher to feed the (green) gas into the NGI and utilize it in large power plants. This would increase the electric energy output from the same gas by ~ 17% and replace fossil natural gas in the NGI at the same time.

Isolated feed-in endeavors of WWTPs are made, e.g., at Hamburg WWTP [39] or Pfaffenhofen WWTP [40]. First efforts of rethinking for plants larger than 30,000 PE_{120} of feeding in the biogas as a suitable alternative to CHP usage on a countrywide scale are made in Switzerland [19,41]. In 2019, a guideline [42] for the plant operator and planner was published, giving specific advice for decision-making and implementations in WWTPs. Both systems got their advantages and drawbacks shown in Table 2. Finally, decision-making will be determined economically and differ widely due to local boundary conditions. Compared to both options, revenues of selling energy (gas, electrical energy and heat) and energy trading on energy exchanges (e.g., control energy and load-management) have to exceed the incurring costs for construction, machinery, energy, maintenance and legal authorizations [42].

Table 2. Advantages and disadvantages of using digestion gas on-site and feeding into the natural gas infrastructure (NGI) (adapted and extended from [42]). WWTP: wastewater treatment plant.

Electrical Energy Production on-Site	Feeding into the NGI
(+) use for self-sufficiency purposes (reduced fees in power purchasing)	(+) full use of the energy content of the digestion gas
(+) heat demand of the WWTP fully covered	(+) substitution of fossil CH_4 in the NGI
(+) local production of renewable energy	(+) separated CO_2 due to the gas treatment usable as a green CO_2 source for PtX systems
(+) usable surplus heat for local district heating	(+) new opportunities for alternative heat concepts
(+) emergency power generation	
(+) providing control energy for energy grids	(−) gas treatment installations required
(−) excess heat possibly not usable	(−) new heat management required
(−) lower efficiency than utilization in bigger power plants	(−) higher costs for electricity purchase due to not fitting taxation systems
(−) no incentives for a utilization of otherrenewable heat sources (e.g., heat recovery)	(−) highly dependent on the local boundary conditions (e.g., connection and distance to the NGI)

(+) Advantage; (−) Drawback.

The waste heat from the CHP units usually covers the heating demand of the digester tanks and the operating buildings. Abandoning the standard of using the CHP waste heat could lead to a higher energy demand for heating purposes due to a feed-in instead of an on-site utilization of the gas. However, this fact may be the missing incentive for plant operators to integrate new renewable heat solutions, like heat recovery from wastewater. Furthermore, the most digestive tanks lack a proper insulation. Improving the reduction of heat losses could lower the thermal energy demand significantly, down to a manageable amount even without CHP units. For conventional WWTPs with fitting feed-in conditions, it is presumed that implementing PtG concepts are a considerable addition to trigger additional synergies and improve the overall efficiency. The authors of [32] showed that it is possible that the missing waste heat can be substituted by the heat production of an electrolyzer and the methanization reactor as well. Additionally, the produced metabolic by-water during the methanization process can be used in the digestion tanks to enhance biogas production [32].

3.1.2. O_2-Utilization: Usage of O_2 From Electrolysis in Wastewater Treatment

Another synergy for PtG in WWTPs is the use of O_2 from the electrolyzer. PtG systems are usually focused on H_2 production, thus O_2 is—as an unintended by-product—usually just vented-off due to a missing application [26]. During the different wastewater treatment steps, O_2 is needed in the aeration tanks for the biological degradation of carbon and nitrogen (nitrification) and can also be utilized to produce ozone (O_3) for micropollutant removal or sludge disintegration.

The intake of O_2 for biological treatment is usually realized by inserting compressed air (blowers/compressors) into the system and supplying the bacteria with the needed oxygen. This part of wastewater treatment is the biggest energy consumer, with 50%–70% of the overall energy consumption [43,44]. Based on that fact, it is often claimed that big energy savings could be achieved due to a reduced gas volume needed, caused by higher O_2 rates (pure O_2 got a five-times higher O_2 content than air). However, there are contradictory scientific evidences for the actual feasibility and impacts of an implementation beyond theoretical considerations. Especially, the benefits of a pure O_2 system in comparison to a conventional municipal system on a large scale are insufficient [45]. Based on a market analysis conducted in [46], using pure O_2 is usually more interesting in treating industry wastewater than for municipal WWTPs.

The major benefits and drawbacks of using pure O_2 instead of compressed air to enhance wastewater treatment are shown in Table 3.

Table 3. Benefits and drawbacks for using pure oxygen for aeration purposes in wastewater treatments.

Benefits	Drawbacks
• Reduction of the needed gas volume due to a five-times higher O_2 content of pure O_2 than compressed air [47] • Better O_2 transfer rate and higher dissolved O_2 concentrations due to higher possible biomass concentrations [48] • Smaller and more compact basin constructions are possible for locations with limited spatial conditions [46] • Better sludge sedimentation is possible, lower specific sludge production and better sludge volume index [49] • Pure O_2 is usable to support aeration processes as an additional system and compensate peaks in O_2 demand caused by, e.g., tourism regime or viticulture [47,50] • Good short/medium-term reconstruction measures for capacity increases in purification quality [51]	• Reduction of the pH level due to insufficient CO_2 stripping, which can inhibit nitrification processes [46,47] • An additional mixing system is required because of the reduced gas flow rates [47,48] • Higher demands on material quality and safety measures [48,52] • Problems with sludge sedimentation could occur [48,51] • Usually, higher operation costs, if no other synergies, are used [32,46,48]

In the following, promising conductions related to enhanced purification quality and/or energy savings are summarized.

One of the first worldwide electrolyzer implementations in WWTPs for O_2 utilization was conducted in 2002 in Barth (Germany) to support the biological treatment during high-load phases and avoid capacity extensions. The annual average load amounts to 10,000 PE_{120}, but during tourism season, the loads peak up to 24,000 PE_{120}, and the plant was not able to handle those peaks with the existing plant configuration. With the electrolyzer-driven supportive O_2 system, the seasonal influence was manageable, and no capacity extension was needed [50,53]. The PEM electrolyzer was used from 2002–2007, and since 2009, an alkaline electrolyzer generates O_2 for treating the wastewater. However, the H_2 utilization path (fuel-cell bus) was abandoned due to two major break-downs and the lack of further funding [53].

Regarding purification quality, approaches for a pure O_2 system in Spain showed no significant increase in COD (chemical oxygen demand), BOD (biochemical oxygen demand) or TN (total nitrogen) removal, whilst energy savings were assumed [54]. In contrast, a study in Germany showed that, in comparison with a conventional system, an additional reduction of up to −20% for COD and TN is possible, but no energy savings could be achieved [47]. At the Nürnberg WWTP (Germany), pure oxygen is successfully used for years as the first biological treatment step for COD removal [55], and at the Lynette WWTP (Copenhagen, Denmark), pure O_2 was used to support the biological treatment step for 15 years. In Neustadt (Germany), a pure oxygen system was implemented and used for years to handle and ensure the legal discharge values due to high viticulture-related load peaks of nearly 100% compared to the usual loads [51].

At the PtG implementation in Avedøre (Denmark), it is shown via laboratory experiments that the use of pure O_2 is possible, and purification processes are not inhibited by the use of pure O_2. Furthermore, the already pressurized O_2 from the electrolyzer is able to displace the corresponding air blower operation to supply the aeration basins [32]. The implementation of an O_2 utilization would increase the overall system economy but was not implemented in full scale due to high investment costs caused by misfitting local plant circumstances. Nevertheless, in combination with an above-described PtG system, the feasibility in WWTPs could be reached towards a single investment (cf. [32]).

For pure oxygen systems, an additional mixing system is required, and deeper basins are favorable due to a longer contact of the O_2 bubbles with the media. The authors of [46,47] showed that shallow basins or volatile fill levels might cause losses in the purification quality. Furthermore, the reduced gas volume of the pure O_2 system is insufficient for mixing the media and preventing sludge settling in common aeration tanks. This also causes a lower CO_2 stripping, which could result in a reduced pH level for wastewater with low buffer/acid capacity and inhibited nitrification processes at pH levels under 6.6. To solve this problem, it could be required to add chemicals (e.g., milk of lime; cf. [47]) or add additional aeration times for the system (cf. [46]) to sustain the biological treatment. This issue could also occur with aeration tank depths of 6 meters or deeper [56]. Other cost savings could thereby be offset, leading to similar or higher total operating costs for the use of pure O_2 systems. Claimed reduced overall basin volume or smaller basins must be viewed critically as well, because sludge concentration is not limited by the O_2 input but of the thickening in the secondary clarifier. Nevertheless, due to the needed deeper basin construction, the space requirements are reduced and favorable for, e.g., very limited field conditions. Whilst [49] claims better sludge characteristics like sedimentation properties or sludge volume indices, in contrast, [48,51] could not observe a similar effect. Additionally, for an implementation of an O_2 system, an expensive O_2 storage and piping infrastructure is needed, with special arrangements regarding corrosion, higher fire risk and a special safety concept due to dealing with pure O_2 [48]. Further instructions regarding handling O_2 in WWTPs are given in [52].

In summary, the results of the evaluated scientific papers are partially contradictory, e.g., enhanced purification quality or space requirements. However, the long-term experiences of several WWTPs with supportive O_2 utilization show that it is even on a large scale possible and reasonable if the local boundary conditions are fitting or require special arrangements. However, no economic statements were given in those publications. The popular claim that using pure O_2 will result in energy (and cost) savings for the WWTP could not be proven in any practical implementation, but higher operating costs were mentioned in [51]. Nevertheless, it can be stated that at least similar or better purification results could be achieved with the use of pure oxygen regarding COD, BOD and NH_4-N (e.g., [47,54]) and energy savings are theoretically possible (e.g., [32]).

3.1.3. O_2 Utilization: Micropollutant Removal via Ozone Produced from Renewable O_2

At present, removing micropollutants is getting more and more important for the water sector. In Switzerland, it is even ordered by law for municipal WWTPs above 80,000 PE_{120} [57]. This is discussed in several other countries as well (e.g., Germany, [58]), or additional steps for micropollutant removal are set up on plants unsolicited (e.g., WWTP Mannheim, Sindelfingen, etc.; Germany). Besides adsorption on activated carbon, oxidation of the micropollutants is a suitable removal process. This leads to another future benefit in operating an electrolyzer in WWTPs by upgrading O_2 to O_3, which could be used in a further treatment step. This additional process stage will result in a substantially higher energy demand for WWTPs, e.g., for a 100,000 PE_{120}, WWTP 22.5 kWh/(PE*a) to 58.3 kWh/(PE*a) (cf. [59]). In a future holistic WWTP concept, O_2 is produced by a RE-driven electrolyzer and closes another loop of resource and overall efficiency. In the next steps, O_3 is produced and the water treated, followed by a biological treatment step to remove the by-products due to the ozonization process to ensure the effluent quality [60]. Using the ozone to remove micropollutants is a commonly used technology in WWTPs [61,62]. Guidelines for the preinvestigation, design and operation are described in, e.g., [63]. This makes ozonization a preferable option for removing micropollutants in combination with PtG systems, leading to more feasibility for both applications. Like dealing with O_2, O_3 requires special precautions in handling as well, which should not be underestimated in daily business. Those are, for instance: use of specific materials, needed O_3 detectors, ventilation systems and safety concepts [64]. First approaches in the form of feasibility studies for a full-scale application for such a system are conducted at WWTP Mainz (Germany) with a planned start in 2021 [65,66].

4. Heat-Sector

PtH describes technologies that transform the electrical surplus energy of RES into heat. The scarce available publications on that topic substantiate that using electrical energy to utilize and store heat in a PtH concept is not an issue and not state-of-the-art in WWTPs so far. The dominant topics in the heat sector are mostly regarding internal optimization concepts using the waste heat for building heating, sludge treatment and in anaerobic sludge digestion (cf. [67]), as well as heat recovery from the sewer system or influent/effluent of the WWTP (cf. [68,69]). Nevertheless, PtH and heat concepts hold a respectable potential for sector coupling, even if they are not in the actual focus for plant operators.

In addition to a high electricity demand, WWTPs with anaerobic sludge digestion usually also have a high demand for heat of 30 to 50 kWh/(PE*d) with big seasonal fluctuations [70]. Compared to electrical energy, heat is an even more complex field to manage. This is caused by the different parameters, which have to match in terms of the amount of needed heating energy, time of demand, different temperature levels, various possible technologies and the nearly unique framework of each WWTP. Heat supply is usually handled as a by-product from the production of electrical energy from the CHP units and, therefore, not demand-driven and in competition with on-site electricity production [70]. Further information in this regard and specific proposals for the implementation of optimized heat and cooling concepts for WWTPs are given in [67].

Typical heat sources in WWTPs are exhaust gases from the CHP units and other combustion processes (e.g., pyrolysis and hydrothermal carbonization); waste heat from aerators/blowers and drying processes or heat recovery from wastewater [71]. Typical heat sinks are given in forms of sludge treatment (e.g., preheating, sludge drying and thermal disintegration); building heating and hot water preparation and heating of the digestion tank [67].

Especially, the digestion tanks are a reasonable heat sink, which is interesting for further investigations. They are usually operated on a mesophilic temperature level of 34 °C to 38 °C; however, a thermophilic temperature level of 48 °C to 55 °C is possible as well [72]. As a biological process, minor fluctuations are not harmful for the system, but effects in biogas production are possible, depending on the rate of increasing the temperature due to a needed adaptation time for the microorganisms. The authors of [73] showed that temperatures up to even 50 °C (change to thermophilic temperature level) are possible without losses in biogas quantity and quality with stable processes in the digestion tank. In this regard, existing studies show different effects concerning methane yield, methane concentrations and needed adaptation time (cf. [73–75]). Thereby, a gradual temperature change is advisable and should not exceed 2 °C per week [76], whilst a direct heating within 24 h could result in a drop of methane yield and methane concentration [74]. Under the approach of using the digestion tanks for PtH purposes, a calculated raise or lowering of the usual operation temperature level is intended, and the digestion tank is used as a "hot water storage". The potential of this storage option is huge: with an average digestion tank volume of 50 l/PE [72] and a specific heat capacity of water of 4.19 kJ/(l*K), the specific energy storage results in 58.2 kWh/(K*PE) or nearly 17.5 GWh for a 100,000 PE_{120} WWTP for a temperature rise of 3 K.

A PtH concept can be implemented directly via heating rods/boilers or indirectly via different types of heat pumps. However, electrode boilers are commonly used for PtH on an industrial level due to the need of a high-temperature process heat. It is possible to integrate such a system into existing heat cycles, but it depends highly on the local boundary conditions. Requirements are an all-the-year heat sink and a power grid connection with sufficient electrical power reserve [9].

Published practical implementations on a scientific level for "classic" PtH concepts in WWTPs are hardly available. Nevertheless, some holistic approaches in local energy concepts demonstrate promising interconnections between WWTPs and the heat sector. The authors of [77] showed for a commercial district located in Milan (Italy) that a PtH system complemented by sewage heat recovery is competitive to individual, distributed heat pumps. In addition, WWTPs are able to participate in local heat supply systems by feeding-in surplus heat [35,70,78,79]. This is realized, e.g., at the Hamburg WWTP, which is feeding heat into the district heating systems for the container terminals of the port of Hamburg. Furthermore, the Hamburg WWTP is testing an aquifer storage by using the groundwater to store heat surplus during the summer to compensate for seasonal fluctuations and save heating energy in the winter [79,80]. The aquifer storage is located beneath the WWTP and includes heat of the nearby industry as well—summing up to nearly 400 GWh/a. By using 100% green energy, this system is able to provide CO_2-free district heating with economically acceptable costs of avoiding CO_2 emissions below 100 € per ton and year [80].

Thermal energy potentials of the Austrian WWTPs are stated in [81], showing that WWTPs are able to contribute significantly in local district heating concepts by recovering energy from digester gas production and wastewater effluent. Based on that, the authors of [78] proposed a set of methods to integrate the potentials into local energy concepts, considering spatial, environmental and economic issues. It is shown that the heat generation from WWTPs offers an alternative to conventional heat generation at competitive costs.

Besides heat, cooling concepts are also an interesting approach to using energy surplus. Cooling is needed in WWTPs for the air conditioning of the operation buildings and cooling down the processed heat (e.g., blowers for aeration and digestion gas) [35]. This can be realized due to, e.g., sorption chillers, which work like heat pumps but just the other way around. Whereas, in heat pumps, a lot of heat is to be dissipated at a higher temperature, in chillers, the aim is to absorb as much heat as possible at a low temperature and, thus, provide cooling [9]. Further information regarding different cooling technologies are given in [82].

Regarding economic feasibility, it can be stated that single-handed PtH systems (even on an industrial level) were in the past, just in combination with high revenues from the control energy market profitable. On the basis of significantly decreasing prices, new business cases have to be found [18]. Nevertheless, compared to other general storage concepts (e.g., batteries, load-shifting, etc.), thermal storages are cheaper, even if prices for other technologies are dropping [9]. However, even if a single implementation of classic PtH systems is not yet feasible, the chances to utilize surplus thermal energy from WWTPs are given but considerably underestimated or unknown. The actual benefit may not be using surplus energy from the subordinate energy grids but taking a more significant role in the local energy system. Integrating WWTPs in local district heating is possible without high investment costs compared to new developments in the vicinity of towns due to an intelligent use of the existing and surrounding infrastructure, provided that the spatial framework is suitable. In addition, thermal energy from WWTPs is a continuous and reliable source of energy, which can be substantially used in district heating, substituting fossil energy sources and reducing greenhouse gas emissions (cf. [78,81]).

5. Mobility Sector and Power-to-Fuel

The mobility sector is one of the most relevant factors and affected by the biggest challenges in reducing greenhouse gas emissions and energy consumption [83,84]. In perspective, it is predicted that, especially, road transport will rise even more and double in terms of CO_2 emissions from 2010 to 2050, up to 14–18 Gt CO_2 [85]. Therefore, a change in transport is mandatory to achieve climate protection goals. In fact, to reduce greenhouse gas emissions, decrease dependencies from crude oil imports and emissions of pollution and noise in cities, the mobility sector has to be nearly completely decarbonized [86]. The potential of RES is huge, e.g., using H_2 in vehicles: 1.0 kWh H_2 is able to replace 1.5 to 2.2 kWh fossil fuel in transport and prevents 465–680 $CO_{2\text{-eq}}$ [9].

To realize defossilization under the use of RES in the transport sector, different types of technologies, types of drive and fuels are available, each with different technology levels and market penetration rates. These include e-mobility, fuel-cell cars and combustion engines using fuel made from renewables. The comparison of the efficiency of drive concepts shows advantages for battery electric vehicles [87]. That is one reason why many car companies actually focus on this concept. Nevertheless, the comparison of total life-cycle greenhouse gas emissions shows that the difference is not as relevant as it seems when only comparing efficiency. All type of drives have advantages and disadvantages, and the assessment of life-cycle greenhouse gas emissions depends heavily on the assumed boundary conditions (e.g., battery size, range and RES) of the individual considered study [88]. Therefore, all types of drive will find its application area and will be part of a mix of technologies in the mobility sector. This variability will be needed to progress in achieving climate protection goals (cf. Table 4).

Table 4. Different types of drives and used renewable fuels [87].

Type of Drive	Type of Fuel	Efficiency of Drive Concept (%)	Life-Cycle Greenhouse Gas Emissions * (g $CO_{2\text{-}eq}$/km)
Battery electric vehicle + battery	Electrical energy	75–85	40–80
Fuel cell + electric vehicle	Hydrogen	25–35	50–70
Combustion engine	Synthetic fuel	10–20	-
Combustion engine	Biofuel	-	-

* [88]: overall driving performance of 150,000 km, battery 60 kWh and further assumptions.

Though, advanced biofuels and biomethane are still represented just in small shares; especially biomethane is growing rapidly in some countries [6].

The production of fuels for the mobility sector in WWTPs is not state-of-the-art but demonstrated in several projects (e.g., [45,89]). Besides providing electrical energy for e-mobility, there are basically two types of fuels usable: gaseous fuels (H_2 and CH_4) and liquid fuels (biofuel and synthetic fuel).

The production of gaseous fuels (CH_4 and H_2) by WWTPs using RES was discussed before in the PtG section of this paper. The utilization of these fuels in the mobility sector is successfully demonstrated in several practical implementations (e.g., [50]). The Henriksdal WWTP in Sweden produces and upgrades biogas for 280 buses used in public transportation [90]. Furthermore, the PtG plant at the Pfaffenhofen WWTP (Germany) will serve as a CH_4-fuel provider for 250 buses as of 2020 [91]. Further small-size mobility studies (1 to 6H_2 buses) are conducted in Barth, Kaisersesch and Sonneberg, Germany [45,50,92]. Another way to produce H_2 is possible via gas reformation from the biogas and was tested in Bottrop (Germany), as well as methanol synthesis (power-to-liquid), but not continued due to missing profitability whilst the technical feasibility could be shown (cf. [89,93]). Another way to produce H_2 is the fermentative conversation of organic mass like sewage sludge or molasses, called "dark fermentation" (cf. [94]).

Besides WWTPs as power-to-mobility locations, there are a lot of ongoing and planned projects given in [21] related to the use of H_2 and CH_4 for transportation in Europe.

WWTPs are able to produce liquid biofuels as well (state-of-research), e.g., through microalgae using wastewater as a nutrient source. This offers a (theoretically) efficient way to remove nutrients than conventional tertiary treatment together with a biomass production that does not use agricultural land or compete with food crops but still lacks feasibility on a large scale [95]. The drawbacks are a high demand of light and a subsequent poor performance during the wintertime for European climatic conditions. This would result in needed technical lighting and big basins, which can offset any feasibility.

Furthermore, the economic production of H_2 and synthetic fuel from sewage sludge is examined in Rosenheim, Germany [96]. Advantages of synthetic liquid fuel and CH_4 (as a natural gas substitute) is the usability in existing and established combustion technologies, which can be used as interim solutions to a future, e.g., H_2-based mobility that differs substantially in the forms of the types of drive and needed infrastructure from the conventional system [97]. WWTPs are able to accompany that transformation by providing the currently required type of fuel and reducing the construction of interim solutions.

At first sight, fuel-based vehicles seem concurrent to electric vehicles. However, the existing energy system cannot provide enough capacity for a full-scale private e-mobility, especially not-controlled charging in local electric distribution grids. Furthermore, battery-driven vehicles are not suitable for transportation vehicles [98]. As a result, a mix of fuel-based mobility for long distances and transportation of goods, as well as electric vehicles for individual mobility in the cities, could be a suitable solution.

Apart from technical issues, the high demand of special raw materials for batteries and fuel cells raise additional ecological problems, which have to be taken into account. The authors of [99] state that tolerable environmental effects, working conditions in mining areas and violations of fundamental rights, as well as the needed amount of raw materials on a large scale, are not given for an abrupt transformation in transportation. Along with technical progress, the conditions of mining, the needed materials and their supply chains have to be developed gradually at the same time. This is transferable to other sectors, like catalysts in the electrolyzer or battery-storage systems.

From a realistic point of view, the capacities of WWTPs compared with the entire demands of fuel are just a contribution to a future solution. However, the authors of [45] showed that the provided H_2 by WWTPs corresponds with the amount of needed fuel in the respective urban area with the size of the WWTP. Therefore, a WWTP of 50,000 PE_{120} that generates O_2 for aeration purposes via a PtG system comprises the H_2 potential of more than an average-sized filling station. The technical proportions between the PtF-concepts and WWTPs fit, but implementations are mostly in combination with other reasonable energy concepts. Nevertheless, this fact is no disadvantage, and WWTPs offer considerable opportunities to take a part in the local sector of mobility as a fuel provider in terms of production and location for filling stations. Combining fuel production and public transport, especially in combination with other synergies, seems reasonable.

6. Conclusions

The evaluation in this paper illustrates the wide spectrum of interactions and possibilities for WWTPs. It is shown that the water sector is capable of taking part in sector coupling in many ways and generates multiple synergies and advantages towards a stand-alone solution of the different applications. The scale of an implementation in WWTPs is mostly much lower than on an industrial level but with a transferability to a high number of suitable locations and individual solutions. For many cases, a quantitative statement from literature/reports is not transferable due to the laboratory scale. Mostly, just the opportunity and feasibility could be evaluated, but in the end, it has to be tested individually on the considered plant. First, practical implementations are realized by small, local networks taking environmental protection as a top priority for their own local policy development, even without funding.

Across all the literature, obstacles and challenges resemble one another. Technological feasibility is mostly given, but the most relevant obstacle is the lack of economic viability driven by a regulatory framework that does not fit to a fast-changing energy environment. That relates, e.g., to Germany, from not fitting the taxation treatment of energy storage systems and power purchases to inconsequential funding mechanisms and other regulatory backgrounds, which lag behind the technological progress for a fast energy transition [6]. There is a need for establishing a framework for sector coupling regarding the special, manifold and complex interactions of the many affected sectors with each other. This includes a further development and coordination of norms and standards across the national level as well.

Furthermore, WWTP operators have to be supported in day-to-day businesses to handle additional tasks caused by sector coupling. Key objectives of the plants will always be treating wastewater as the main priority. Especially, smaller WWTPs are often not able to provide the needed financial and human resources for additional activities in such a variety of tasks outside the usual disciplines. New business models for operating innovative technologies have to be found and to make use of as much of the synergies as possible. WWTPs would switch from locations of just treating wastewater to local energy and resource facilities, with all their benefits and drawbacks. In conclusion, the water sector is a notable and reasonable player in the field of sector coupling, especially in times of changes to a renewable energy system.

7. Outlook and Further Approach

To implement such innovative technologies in WWTPs, a holistic analysis of the available resources on the plant is necessary, e.g., energy production, flexibility options and gas flow rates. These have to match with the surrounding infrastructure (industry, public transportation system and energy grids) and the energetic environment (local heat demands, available energy surplus, etc.). For the Mainz WWTP (400,000 PE_{120}; Germany), a wide-range sector-coupling analysis was carried out, which includes an overall analysis followed by a detailed feasibility study for the most promising results [65,66]. In the following, based on these conductions, the findings that are necessary to prepare and realize such an implantation are presented:

- The upcoming needed technical knowledge of every sector is very diverse and causes many interconnections. This requires interdisciplinary expertise, with challenging tasks for every specialist in his field, which cannot be handled by a single player.
- Whilst every expert group can handle their specific field the best, the interconnections can rise up to major problems due to missing experiences and require a very close cooperation.
- Coordination and handling of the utilization path among each other of the competing available resources and their data management is very challenging in terms of gas flow rates, energy production, heat demands and storage management.
- Linking new components with significant impacts on plant processes to existing systems is difficult, e.g., coordination of the electrolysis gases: O_2 production, O_3 generation and O_3 demand for wastewater treatment or the use of H_2 by supplying public transportation and feeding into the NGI.
- The plant operators and the political decision-makers have to be involved and convinced from the start to carry out and promote such big municipal implementations and investments.
- Only a motivated and reliable team is able to keep up with a challenging and difficult regulatory environment.
- Local and/or federal funding is up-to-now mandatory to use the potential of sector coupling in WWTPs.

Assembling a responsible interdisciplinary team is therefore a prerequisite for working on such a diversified project. This includes at least one of the following experts: general planner for large construction measures, plant operator, energy supplier, grid operator, scientific support and various engineering expertise depending on the targeted sectors. This team should develop a practicable and technical-economic feasible concept based on the local situation. After selecting and specifying a preferred variant to be implemented, the funding situation needs to be clarified before finally deciding on the implementation.

In the meantime, at the Mainz WWTP, the political and technical decision-maker has given their consent, and the positive notification of federal funding has arrived. The project will start in 2020 with further regulatory preparations, assessments and the invitation of tenders. The start of implementation is planned for 2021.

Author Contributions: Conceptualization, M.S.; methodology, M.S.; formal analysis, M.S. and O.G.; investigation, M.S. and O.G.; writing—original draft preparation, M.S.; writing—review and editing, M.S., O.G. and H.S.; supervision, H.S.; All authors have read and agreed to the published version of the manuscript.

Funding: This research received no external funding.

Conflicts of Interest: The authors declare no conflict of interest.

References

1. Friedrich, E.; Pillay, S.; Buckley, C.A. Environmental life cycle assessments for water treatment processes—A South African case study of an urban water cycle. *Water SA* **2009**, *2009*, 73–84. [CrossRef]
2. U.S. Department of Energy. The Water-Energy Nexus: Challenges and Opportunities. 2014. Available online: https://www.energy.gov/downloads/water-energy-nexus-challenges-and-opportunities (accessed on 13 March 2020).
3. Beca Consultants Pty Ltd (Beca). Opportunities for Renewable Energy in the Australian Water Sector. 2015. Available online: https://arena.gov.au/assets/2016/01/Opportunities-for-renewable-energy-in-the-Australian-water-sector.pdf (accessed on 13 March 2020).

4. Economic and Social Council, United Nations. Special Edition: Progress Towards the Sustainable Development Goals. 2019. Available online: https://undocs.org/E/2019/68 (accessed on 13 March 2020).
5. UNFCCC. Paris Agreement. 2016. Available online: https://unfccc.int/files/meetings/paris_nov_2015/application/pdf/paris_agreement_english_.pdf (accessed on 13 March 2020).
6. Renewable Energy Policy Network for the 21st Century (REN21). Renewables 2019—Global Status Report. 2019. Available online: https://www.ren21.net/wp-content/uploads/2019/05/gsr_2019_full_report_en.pdf (accessed on 13 March 2020).
7. BNetzA. Flexibilität im Stromversorgungssystem. Bestandsaufnahme, Hemmnisse und Ansätze zur Verbesserten Erschließung von Flexibilität. 2017. Available online: https://www.bundesnetzagentur.de/SharedDocs/Downloads/DE/Sachgebiete/Energie/Unternehmen_Institutionen/NetzentwicklungUndSmartGrid/BNetzA_Flexibilitaetspapier.pdf?__blob=publicationFile&v=1 (accessed on 13 March 2020).
8. Deutsche Energie-Agentur GmbH (DENA). dena-Netzflexstudie. Optimierter Einsatz von Speichern für Netz- und Marktanwendungen in der Stromversorgung. 2017. Available online: https://www.dena.de/fileadmin/dena/Dokumente/Pdf/9191_dena_Netzflexstudie.pdf (accessed on 13 March 2020).
9. Sterner, M.; Stadler, I. *Energiespeicher—Bedarf, Technologien, Integration*; 2.Auflage 2016; Springer: Berlin, Germany, 2016; ISBN 978-3-642-37379-4.
10. Müller-Syring, G.; Henel, M.; Köppel, W.; Sterner, M.; Höcher, T. Entwicklung von Modularen Konzepten zur Erzeugung, Speicherung und Einspeisung von Wasserstoff und Methan ins Erdgasnetz. 2013. Available online: https://www.dvgw.de/medien/dvgw/forschung/berichte/g1_07_10.pdf (accessed on 15 April 2020).
11. Lensch, D.; Schaum, C.; Cornel, P.; Lensch, D.; Schaum, C.; Cornel, P. Examination of food waste co-digestion to manage the peak in energy demand at wastewater treatment plants. *Water Sci. Technol.* **2016**, *73*, 588–596. [CrossRef] [PubMed]
12. Seier, M.; Schebek, L. Model-based investigation of residual load smoothing through dynamic electricity purchase: The case of wastewater treatment plants in Germany. *Appl. Energy* **2017**, *205*, 210–224. [CrossRef]
13. Mauky, E.; Weinrich, S.; Jacobi, H.-F.; Nägele, H.-J.; Liebetrau, J.; Nelles, M. Demand-driven biogas production by flexible feeding in full-scale—Process stability and flexibility potentials. *Anaerobe* **2017**, *46*, 86–95. [CrossRef] [PubMed]
14. Schäfer, M.; Hobus, I.; Schmitt, T.G. Energetic flexibility on wastewater treatment plants. *Water Sci. Technol.* **2017**, *76*, 1225–1233. [CrossRef]
15. Schmitt, T.G.; Schäfer, M.; Gretzschel, O.; Knerr, H.; Hüesker, F.; Kornrumpf, T.; Zdrallek, M.; Salomon, D.; Bidlingmaier, A.; Simon, R.; et al. Abwasserreinigungsanlagen als Regelbaustein in intelligenten Verteilnetzen mit Erneuerbarer Energieerzeugung—arrivee. 2017. Available online: www.erwas-arrivee.de (accessed on 25 June 2018).
16. Schäfer, M. Ein Methodischer Ansatz zur Bereitstellung Energetischer Flexibilität Durch Einen Anpassungsfähigen Kläranlagenbetrieb. Ph.D. Thesis, Technische Universität Kaiserslautern, Kaiserslautern, Germany, February 2019. Available online: https://nbn-resolving.org/urn:nbn:de:hbz:386-kluedo-56084 (accessed on 13 March 2020).
17. ASUE. Power to Gas. Erzeugung von Regenerativem Erdgas. 2014. Available online: https://asue.de/sites/default/files/asue/themen/umwelt_klimaschutz/2014/broschueren/07_06_14_Power_to_Gas.pdf (accessed on 13 March 2020).
18. Forschungsstellefür Energiewirtschaft e.V. (FfE). Kurzstudie Power-to-X. Ermittlung des Potenzials von PtX-Anwendungen für die Netzplanung der Deutschen ÜNB. 2017. Available online: https://www.netzentwicklungsplan.de/de/ermittlung-des-potenzials-von-power-x-anwendungen-fuer-die-netzplanung-zu-kapitel-251 (accessed on 13 March 2020).
19. Peyer, T.; Rene, N.; Thomas, H.; Monika, R. Kläranlagen-Ideal für Power-to-Gas: Swisspower identifiziert 100 Kläranlagen in der Nähe von Gasnetzen. *AQUA GAS* **2016**, *7–8*, 42–46.
20. Schäfer, M.; Gretzschel, O.; Schütz, S.; Schuhmann, E.; Raabe, T. The natural gas grid infrastructure as a suitable storage for renewable energy produced by wastewater treatment plants. In Proceedings of the 10th International Renewable Energy Storage Conference (IRES), Düsseldorf, Germany, 15–16 March 2016.
21. Wulf, C.; Linßen, J.; Zapp, P. Review of Power-to-Gas Projects in Europe. *Energy Procedia* **2018**, *155*, 367–378. [CrossRef]

22. Schäfer, M.; Gretzschel, O.; Schmitt, T.G.; Knerr, H. Wastewater Treatment Plants as System Service Provider for Renewable Energy Storage and Control Energy in Virtual Power Plants—A Potential Analysis. *Energy Procedia* **2015**, *73*, 87–93. [CrossRef]

23. Hien, S. Approaches for Supportive Prediction of Biogas Production Rate and Control Strategies to Provide Flexible Power Production. Ph.D. Thesis, University of Luxembourg, Esch-sur-Alzette, Luxemburg, May 2017. Available online: http://hdl.handle.net/10993/31245 (accessed on 13 March 2020).

24. Smolinka, T.; Günther, M.; Garche, J. Stand und Entwicklungspotenzial der Wasserelektrolyse zur Herstellung von Wasserstoff aus regenerativen Energien: NOW-Studie: Kurzfassung des Abschlussberichts—Redakionsstand: 22.12.2010 (Revision 1 vom 05.07.2011). 2011. Available online: https://www.tib.eu/de/suchen/id/TIBKAT%3A872387518/Stand-und-Entwicklungspotenzial-der-Wasserelektrolyse/ (accessed on 16 March 2020).

25. DENA. *Power to Gas. Eine Innovative Systemlösung auf dem Weg zur Marktreife*; Deutsche Energie-Agentur GmbH, DENA: Berlin, Germany, 2013.

26. Graf, F.; Götz, M.; Henel, M.; Schaaf, T.; Tichler, R. Technoökonomische Studie von Power-to-Gas-Konzepten. 2014. Available online: https://www.dvgw.de/medien/dvgw/forschung/berichte/g3_01_12_tp_b_d.pdf (accessed on 16 March 2020).

27. DWA. *DWA-M 363—Herkunft, Aufbereitung und Verwertung von Biogasen*; DWA, Ed.; DWA: Hennef, Germany, 2010; ISBN 978-3-941897-52-6.

28. Graf, F.; Krajete, A.; Schmack, U. Techno-ökonomische Studie zur Biologischen Methanisierung bei Power-to-Gas-Konzepten. 2014. Available online: https://www.dvgw.de/themen/forschung-und-innovation/forschungsprojekte/dvgw-forschungsbericht-g-30113/ (accessed on 16 March 2020).

29. Reuter, M. Power-to-Gas: Biological methanization; first at a municipal sewage plant. In Proceedings of the 8th International Renewable Energy Storage Conference, Düsseldorf, Germany, 18–20 November 2013.

30. Dröge, S.; Pacan, B. Erfahrungen mit der Power-to-Gas Pilotanlage im Energiepark Pirmasens-Winzeln. In Proceedings of the Fachgespräch, Biologische Methanisierung, Berlin, Germany, 25 April 2017.

31. Trautmann, N.; Nelting, K.; Vogel, B.; Weichgrebe, D.; Stopp, P.; Cuff, G. Methan aus Erneuerbaren Energien—Biologische Umwandlung von Wasserstoff aus der Elektrolyse zu Methan. 2018. Available online: https://www.dbu.de/OPAC/ab/DBU-Abschlussbericht-AZ-33505-01.pdf (accessed on 16 March 2020).

32. Lardon, L.; Thorberg, D.; Krosgaard, L. Biogas valorization and efficient energy management—Technical and economic analysis of biological methanation. In *Powerstep: Your Flush, Our Energy*; Kompetenzzentrum Wasser Berlin: Berlin, Germany, 2018; Available online: http://powerstep.eu/system/files/generated/files/resource/d3-2-technical-and-economic-analysis-of-biological-methanationdeliverable.pdf (accessed on 16 March 2020).

33. Gandiglio, M.; Lanzini, A.; Soto, A.; Leone, P.; Santarelli, M. Enhancing the Energy Efficiency of Wastewater Treatment Plants through Co-digestion and Fuel Cell Systems. *Front. Environ. Sci.* **2017**, *5*, 221. [CrossRef]

34. Wang, H.; Gu, Y.; Li, Y.; Li, X.; Luo, P.; Robinson, Z.P.; Wang, X.; Wu, J.; Li, F. The feasibility and challenges of energy self-sufficient wastewater treatment plants. *Appl. Energy* **2017**, *204*, 1463–1475. [CrossRef]

35. Pinnekamp, J.; Schröder, M.; Bolle, F.-W.; Gramlich, E.; Gredigk-Hoffmann, S.; Koenen, S.; Loderhose, M.; Miethig, S.; Ooms, K.; Riße, H.; et al. Energie in Abwasseranlagen. Handbuch NRW. 2018. Available online: https://www.umwelt.nrw.de/fileadmin/redaktion/Broschueren/energie_abwasseranlagen.pdf (accessed on 16 March 2020).

36. Gretzschel, O.; Schmitt, T.G.; Hansen, J.; Siekmann, K.; Jakob, J. Sludge digestion instead of aerobic stabilisation—A cost benefit analysis based on experiences in Germany. *Water Sci. Technol.* **2014**, *69*, 430–437. [CrossRef]

37. ASUE. *BHKW Kennzahlen 2014/2015. Module, Anbieter, Kosten. Arbeitsgemeinschaft für Sparsamen und Umweltfreundlichen Energieverbrauch e.V.*; ASUE: Berlin, Germany, 2014.

38. Strauss, K. *Kraftwerkstechnik. Zur Nutzung Fossiler, Nuklearer und Regenerativer Energiequellen, 7. Auflage*; Springer Vieweg: Berlin/Heidelberg, Germany, 2016; ISBN 978-3-662-53029-0. [CrossRef]

39. Erbe, V.; Kolisch, G.; Feldmann, N. Studie zur Aufbereitung und Einspeisung von Faulgas auf Kommunalen Kläranlagen. 2011. Available online: https://www.lanuv.nrw.de/landesamt/forschungsvorhaben/sonstiges?tx_cartproducts_products%5Bproduct%5D=741&cHash=baebeccab6c9a99fb823b8f435fc00a2 (accessed on 16 March 2020).

40. Electrochea. Electrochea Realisiert Power-to-Gas-Anlage für ein Nachhaltiges Pfaffenhofen. 2017. Available online: http://www.electrochaea.com/wp-content/uploads/2017/11/20171113_PM-Electrochaea_PtoG_fuer_Pfaffenhofen_DE_FIN.pdf (accessed on 16 March 2020).

41. Zutter, R.; Nijsen, R.; Peyer, T. Studie Potential zur Effizienzsteigerung in Kläranlagen mittels Einspeisung oder Verstromung des Klärgases. 2015. Available online: https://swisspower.ch/themen-und-standpunkte/potential-zur-effizienzsteigerung-in-kl%C3%A4ranlagen-mittels-einspeisung-oder-verstromung-des-kl%C3%A4rgases (accessed on 16 March 2020).

42. Ryser Ingenieure AG. Klärgas-Verstromung oder Aufbereitung und Einspeisung. Entscheidhilfe für Betreiber und Planer. 2019. Available online: http://www.infrawatt.ch/sites/default/files/2019_02_20_Energie%20in%20ARA_Kap._Kl%C3%A4rgas%20Verstromen%20oder%20Einspeisen.pdf (accessed on 16 March 2020).

43. Kolisch, G.; Taudien, Y.; Osthoff, T. Projekt Nr. 2: Verbesserung der Klärgasnutzung, Steigerung der Energieausbeute auf kommunalen Kläranlagen (Zusatzbericht). 2014. Available online: https://www.lanuv.nrw.de/fileadmin/forschung/wasser/klaeranlage_abwasser/2014_Abschlussbericht_TP2.pdf (accessed on 16 March 2020).

44. Maktabifard, M.; Zaborowska, E.; Makinia, J. Achieving energy neutrality in wastewater treatment plants through energy savings and enhancing renewable energy production. *Rev. Environ. Sci. Biotechnol.* **2018**, *17*, 655–689. [CrossRef]

45. Jentsch, M.; Büttner, S. Dezentrale Umsetzung der Energie- und Verkehrswende mit Waserstoffsystemen auf Kläranlagen. *gwf Gas + Energie* **2019**, *6*, 28–39.

46. Rudolph, K.-U.; Müller-Czygan, G.; Bombeck, M. Reinsauerstoffbelüftung auf kleinen Industriekläranlagen—Energieeinsparpotenziale und Kapazitätssteigerungen am Beispiel der Kläranlage der Fa. Emsland Frischgeflügel GmbH: Meschede, Germany. Available online: https://www.dbu.de/OPAC/ab/DBU-Abschlussbericht-AZ-26353.pdf (accessed on 15 April 2020).

47. Büttner, S.; Jentsch, M.; Hörnlein, S.; Hubner, B. Sektorenkopplung im Rahmen der Energiewende—Einsatz von Elektrolysesauerstoff auf kommunalen Kläranlagen. In *Nutzung Regenerativer Energiequellen und Wasserstofftechnik 2018*; Luschtinetz, T., Lehmann, J., Eds.; Fachhochschule Stralsund: Stralsund, Germany, 2018; pp. 22–41. ISBN 978-3-9817740-4-7. Available online: https://www.hochschule-stralsund.de/storages/hs-stralsund/FAK_ETI/Dateien/REGWA/TagungsBaende/Tagungsband_2018-11-04.pdf (accessed on 16 March 2020).

48. CETAQUA. Greenlysis—Final Report. 2013. Available online: https://ec.europa.eu/environment/life/project/Projects/index.cfm?fuseaction=home.showFile&rep=file&fil=LIFE08_ENV_E_000118_AfterLIFE.pdf (accessed on 16 March 2020).

49. Schmid-Schmieder, V. Alternative Energiequellen auf Kläranlagen: Elektrolytische Wasserstoff- und Sauerstoffnutzung auf Kläranlagen—Wasserstoffproduktion aus Faul- und Biogas. In *wwt Wasserwirtschaft Wassertechnik*; HUSS-Verlag: Munich, Germany, 2007; Volume 11–12.

50. Haas, F.; Jain, A.; Lehmann, J.; Luschtinetz, O.; Schefler, R. The hydrogen-oxygen project in Barth. *Int. J. Hydrog. Energy* **2004**, *30*, 555–557. [CrossRef]

51. Hansen, J.; Steinmetz, H.; Zettl, U. Betriebsergebnisse zum Einsatz der Reinsauerstoffbegasung zur weitergehenden Stickstoffelimination bei einer Anlage mit Weinbaueinfluß. *Abwassertechnik* **1996**, *2*, 32–36.

52. Margot, J.; Urfer, D. Sicherheitsaspekte zum Umgang mit Sauerstoff auf Kläranlagen. 2016. Available online: https://www.micropoll.ch/fileadmin/user_upload/Redaktion/Dokumente/02_Faktenblaetter/Faktenblatt_Sauerstoff_DE_FINAL_21112016.pdf (accessed on 16 March 2020).

53. Strässle, M. Anwendungspotentiale der Wasserstofftechnologie auf Kläranlagen. Master's Thesis, Leibnitz University Hannover, Hannover, Germany, October 2017.

54. CETAQUA. GREENLYSIS—Hydrogen and Oxygen Production via Electrolysis Powered by Renewable Energies to Reduce Environmental Footprint of a WWTP 2012. Available online: http://ec.europa.eu/environment/life/project/Projects/index.cfm?fuseaction=search.dspPage&n_proj_id=3416 (accessed on 16 March 2020).

55. SUN. Das Klärwerk 1 in Nürnberg. Eine Kurzbeschreibung. 2017. Available online: https://www.nuernberg.de/imperia/md/sun/dokumente/sun/kw1_kurz.pdf (accessed on 16 March 2020).

56. DWA. *DWA-A 131—Bemessung von Einstufigen Belebungsanlagen, Juni 2016*; Deutsche Vereinigung für Wasserwirtschaft Abwasser und Abfall: Hennef, Germany, 2016; ISBN 978-3-88721-331-2.

57. Gewässerschutzverordnung. Schweizerische Bundesrat. GSchV, 2018. Law Text (AS 1998 2863). Available online: https://www.admin.ch/opc/de/classified-compilation/19983281/index.html (accessed on 28 September 2018).

58. Hillenbrand, T.; Tettenborn, F. Empfehlungen des Stakeholder-Dialogs "Spurenstoffstrategie des Bundes". An die Politik zur Reduktion von Spurenstoffeinträgen in die Gewässer. 2017. Available online: https://www.bmu.de/fileadmin/Daten_BMU/Download_PDF/Binnengewaesser/spurenstoffstrategie_policy_paper_bf.pdf (accessed on 16 March 2020).

59. Pinnekamp, J.; Bolle, F.-W.; Palmowski, L.; Veltmann, K.; Mousel, D.; Mauer, C.; Eckers, S. Energiebedarf von Verfahren zur Elimination von organischen Spurenstoffen. Final Report. 2011. Available online: https://www.lanuv.nrw.de/fileadmin/lanuv/wasser/abwasser/forschung/pdf/Abschlussbericht_ENVELOS.pdf (accessed on 20 March 2020).

60. Schäfer, M.; Schmitt, T.G.; Gretzschel, O.; Steinmetz, H. Integration of fluctuating Renewable Energies on WWTPs to remove micropollutants due ozonation. In Proceedings of the 12th International Renewable Energy Storage Conference—IRES 2018, Düsseldorf, Germany, 13–15 March 2018.

61. UBA. Maßnahmen zur Verminderung des Eintrages von Mikroschadstoffen in die Gewässer 85/2014. 2014. Available online: https://www.umweltbundesamt.de/sites/default/files/medien/378/publikationen/texte_85_2014_massnahmen_zur_verminderung_des_eintrages_von_mikroschadstoffen_in_die_gewaesser_0.pdf (accessed on 16 March 2020).

62. DWA. *Möglichkeiten der Elimination von Anthropogenen Spurenstoffen*; DWA, Ed.; DWA: Hennef, Germany, 2015; ISBN 978-3-88721-210-0.

63. Wunderlin, P.; Abegglen, C.; Durisch-Kaiser, E.; Götz, C.; Joss, A.; Kienle, C.; Langer, M.; Peter, A.; Santiago, S.; Soltermann, F.; et al. Abklärungen Verfahrenseignung Ozonung. Available online: https://www.micropoll.ch/fileadmin/user_upload/Redaktion/Dokumente/03_Vollzugshilfen/Abkl%C3%A4rungenVerfahrenseignungOzonung_DE_FINAL_20042017.pdf (accessed on 16 March 2020).

64. Margot, J.; Urfer, D. Sicherheitsaspekte zum Umgang mit Ozon auf Kläranlagen. 2016. Available online: https://www.micropoll.ch/fileadmin/user_upload/Redaktion/Dokumente/02_Faktenblaetter/Faktenblatt_Ozon_DE_Final_09092016.pdf (accessed on 16 March 2020).

65. Steinmetz, H.; Schmitt, T.G.; Schäfer, M.; Gretzschel, O.; Krieger, S.; Alt, K.; Zydorczyk, S.; Bender, V.; Pick, E. Konzeptstudie. Klimafreundliche und ressourceneffiziente Anwendung der Wasserelektrolyse zur Erzeugung von regenerativen Speichergasen kombiniert mit einer weitergehenden Abwasserbehandlung zur Mikroschadstoffelimination auf Kläranlagen. Unpublished work. 2018.

66. Steinmetz, H.; Schmitt, T.G.; Schäfer, M.; Gretzschel, O.; Krieger, S.; Alt, K.; Zydorczyk, S.; Bender, V.; Pick, E. Ergänzende Betrachtungen zur Konzeptstudie. Unpublished work. 2019.

67. DWA KEK-10.4. *Wärme- und Kältekonzepte auf Kläranlagen: Arbeitsbericht der DWA-Arbeitsgruppe 10.4 "Wärme- und Kältekonzepte auf Kläranlagen"*; Arbeitsbericht. Korrespondenz Abwasser, Abfall; DWA: Hennef, Germany, 2016; pp. 704–713.

68. AWEL—Amt für Abfall, wasser, Energie und Luft. Baudirektion Kanton Zürich. Heizen und Kühlen mit Abwasser. Leitfaden für die Planung, Bewilligung und Realisierung von Anlagen zur Abwasserenergienutzung. 2010. Available online: https://awel.zh.ch/internet/baudirektion/awel/de/energie_radioaktive_abfaelle/waermenutzung_ausuntergrundwasser/abwasser/_jcr_content/contentPar/downloadlist/downloaditems/1596_1433140855118.spooler.download.1433141005406.pdf/Heizen_Kuehlen_Abwasser.pdf (accessed on 16 March 2020).

69. Alekseiko, L.N.; Slesarenko, V.V.; Yudakov, A.A. Combination of wastewater treatment plants and heat pumps. *Pac. Sci. Rev.* **2014**, *16*, 36–39. [CrossRef]

70. Mitsdoerffer, R. Wärme- und Kältekonzepte auf Kläranlagen, München. 2017. Available online: https://www.gfm-ingenieure.de/fileadmin/Daten/Referenzen/Klaeranlage/Energie/Vortrag-Mitsdoerffer-20170115-mi.pdf (accessed on 16 March 2020).

71. DWA. *DWA-M 114—Abwasserwärmenutzung. Merkblatt, Entwurf September 2018*; Deutsche Vereinigung für Wasserwirtschaft Abwasser und Abfall: Hennef, Germany, 2018; ISBN 978-3-88721-635-1.

72. Seyfried, C.F.; Kroiss, H.; Rosenwinkel, K.-H.; Dichtl, N.; Weiland, P. *Anaerobtechnik. Abwasser-, Schlamm- und Reststoffbehandlung, Biogasgewinnung*; 3., neu bearbeitete Auflage; Springer Vieweg: Berlin, Germany, 2015; ISBN 978-3-642-24895-5.

73. Hubert, C.; Steiniger, B.; Schaum, C.; Michel, M.; Spallek, M. Variation of the digester temperature in the annual cycle—Using the digester as heat storage. *Water Pract. Technol.* **2019**, *14*, 471–481. [CrossRef]

74. Bousková, A.; Dohányos, M.; Schmidt, J.E.; Angelidaki, I. Strategies for changing temperature from mesophilic to thermophilic conditions in anaerobic CSTR reactors treating sewage sludge. *Water Res.* **2005**, *39*, 1481–1488. [CrossRef]

75. Barrington, S.; Ortega, L.; Guiot, S.R. Thermophilic adaptation of a mesophilic anaerobic sludge for food waste treatment. *J. Environ. Manage.* **2007**, *88*, 517–525. [CrossRef]

76. DWA. *DWA-M 368—Biologische Stabilisierung von Klärschlamm*; DWA, Ed.; DWA: Hennef, Germany, 2014; ISBN 978-3-944328-60-7.

77. Aprile, M.; Scoccia, R.; Dénarié, A.; Kiss, P.; Dombrovszky, M.; Gwerder, D.; Schuetz, P.; Elguezabal, P.; Arregi, B. District power-to-heat/cool complemented by sewage heat recovery. *Energies* **2019**, *12*, 364. [CrossRef]

78. Kollmann, R.; Neugebauer, G.; Kretschmer, F.; Truger, B.; Kindermann, H.; Stoeglehner, G.; Ertl, T.; Narodoslawsky, M. Renewable energy from wastewater—Practical aspects of integrating a wastewater treatment plant into local energy supply concepts. *J. Clean. Prod.* **2016**, *155*, 119–129. [CrossRef]

79. Hamburg-Wasser. Unterirdischer Waermespeicher Erfolgreich Getestet. 2017. Available online: https://www.hamburgwasser.de/privatkunden/unternehmen/presse/unterirdischer-waermespeicher-erfolgreich-getestet/ (accessed on 16 March 2020).

80. Hansen, G.; Giese, T. "Mach 3": Umweltwärme aus Wasser für eine erneuerbare Fernwärmeversorgung in Hamburg—ein Projekt mit vielen Möglichkeiten. In Proceedings of the 31th Hamburger Kolloquium zur Abwasserwirtschaft: Themenschwerpunkte: Mikroschadstoffe, Mikroplastik, Antibiotikaresistenzen; Industrieabwasserbehandlung; Niederschlagswasser, Umweltwärme; Klärschlammentsorgung, Gewässerschutz, Hamburg, Germany, 18–19 September 2019; Available online: https://cgi.tu-harburg.de/~{}awwweb/downloads/TagungsbandAbwasserkolloquium2019.pdf (accessed on 16 March 2020).

81. Neugebauer, G.; Kretschmer, F.; Kollmann, R.; Narodoslawsky, M.; Ertl, T.; Stoeglehner, G. Mapping Thermal Energy Resource Potentials from Wastewater Treatment Plants. *Sustainability* **2015**, *7*, 12988–13010. [CrossRef]

82. BHKW-Infozentrum. Grundlagen der Kraft-Wärme-Kälte-Kopplung. Available online: https://www.bhkw-infozentrum.de/allgemeine-erlaeuterungen-bhkw-kwk/kwkk-grundlagen.html (accessed on 6 April 2020).

83. IPCC. Climate change 2014. Synthesis Report. Available online: https://archive.ipcc.ch/pdf/assessment-report/ar5/syr/SYR_AR5_FINAL_full_wcover.pdf (accessed on 16 March 2020).

84. IEA. World Energy Outlook. 2018. Available online: https://webstore.iea.org/download/summary/190?fileName=English-WEO-2018-ES.pdf (accessed on 16 March 2020).

85. Pietzcker, R.C.; Longden, T.; Chen, W.; Fu, S.; Kriegler, E.; Kyle, P.; Luderer, G. Long-term transport energy demand and climate policy: Alternative visions on transport decarbonization in energy-economy models. *Energy* **2014**, *64*, 95–108. [CrossRef]

86. Bundesministerium für Verkehr und digitale Infrastruktur (BMVI). Initiative klimafreundlicher Straßengüterverkehr. Fahrplan für einen klimafreundlichen Straßengüterverkehr. 2017. Available online: https://www.bmvi.de/SharedDocs/DE/Anlage/G/MKS/initiative-klimafreundlicher-strassengueterverkehr.pdf?__blob=publicationFile (accessed on 16 March 2020).

87. Unnerstall, T. Verkehr—Die Zukunft des Autos. In *Energiewende verstehen: Die Zukunft von Autoverkehr, Heizen und Strompreisen*; Unnerstall, T., Ed.; Springer: Berlin, Germany, 2018; pp. 57–69. ISBN 978-3-662-57786-8.

88. Sternberg, A.; Hank, C.; Hebling, C. Treibhausgas-Emissionen für Batterie- und Brennstoffzellenfahrzeuge mit Reichweiten über 300 km: Studie im Auftrag der H2 Mobility. Available online: https://www.ise.fraunhofer.de/content/dam/ise/de/documents/news/2019/ISE_Ergebnisse_Studie_Treibhausgasemissionen.pdf (accessed on 11 March 2020).

89. Bolle, F.-W.; Genzowsky, K.; Gredigk-Hoffmann, S.; Reinders, M.; Riße, H.; Schröder, M.; Manja, S.; Wöffen, B.; Illing, F.; Jagemann, P.; et al. WaStraK NRW "Einsatz der Wasserstofftechnologie in der Abwasserbeseitigung"—Phase I. Band I: Kompendium Wasserstoff. 2012. Available online: https://www.lanuv.nrw.de/landesamt/forschungsvorhaben/klaeranlage-abwasserbeseitigung?tx_cartproducts_products%5Bproduct%5D=654&cHash=f6ef363dd1a75575207fe2fa142d5e95 (accessed on 16 March 2020).

90. Scandinavian Biogas. Henriksdal and Bromma, Sweden. Available online: http://scandinavianbiogas.com/en/project/henriksdal-and-bromma/ (accessed on 16 March 2020).
91. Schattenhofer, S. Pfaffenhofen gibt Gas—erneuerbar: Neues Projekt der Bürger-Energiegenossenschaft: Überschüssiger Öko-Strom wird Biomethan. 2017. Available online: https://www.donaukurier.de/nachrichten/wirtschaft/lokalewirtschaft/Pfaffenhofen-Pfaffenhofen-gibt-Gas-erneuerbar;art1735,3569895 (accessed on 16 March 2020).
92. Meier, B. Kläranlage Kaisersesch soll Wasserstofftankstelle speisen. Rhein-Zeitung. 2017. Available online: https://www.rhein-zeitung.de/region/aus-den-lokalredaktionen/kreis-cochem-zell_artikel,-klaeranlage-kaisersesch-soll-wasserstofftankstelle-speisen-_arid,1606117.html (accessed on 29 January 2018).
93. Bolle, F.-W.; Reinders, M.; Riße, H.; Schröder, M.; Bernhard, W.; Illing, F. WaStraK NRW "Einsatz der Wasserstofftechnologie in der Abwasserbeseitigung"—Phase I. Band II: Methanolsynthese. 2012. Available online: https://www.lanuv.nrw.de/landesamt/forschungsvorhaben/klaeranlage-abwasserbeseitigung?tx_cartproducts_products%5Bproduct%5D=654&cHash=f6ef363dd1a75575207fe2fa142d5e95 (accessed on 16 March 2020).
94. Mariakakis, I. A Two Stage Process for Hydrogen and Methane Production by the Fermentation of Molasses. Ph.D. Thesis, Stuttgarter Berichte zur Siedlungswasserwirtschaft, Stuttgart, Germany, 2013.
95. Chen, G.; Zhao, L.; Qi, Y. Enhancing the productivity of microalgae cultivated in wastewater toward biofuel production: A critical review. *Appl. Energy* **2015**, *137*, 282–291. [CrossRef]
96. 2synfuel. Turning Sweage Sluidge into Fuels and Hydrogen. 2018. Available online: http://www.tosynfuel.eu (accessed on 16 March 2020).
97. Energbieangentor.NRW Wasserstoff—Schlüssel zur Energiewende. Beispiele aus Nordrhein-Westfalen von der Herstellung bis zur Nutzung. 2018. Available online: https://broschueren.nordrheinwestfalendirekt.de/broschuerenservice/energieagentur/wasserstoff-schluessel-zur-energiewende-beispiele-aus-nordrhein-westfalen-von-der-herstellung-bis-zur-nutzung/2833 (accessed on 16 March 2020).
98. Wietschel, M.; Plötz, P.; Pfluger, B.; Klobasa, M.; Eßer, A.; Haendel, M.; Müller-Kirchenbauer, J.; Kochems, J.; Hermann, L.; Grosse, B.; et al. *Sektorkopplung—Definition, Chancen und Herausforderungen. Working Paper Sustainability and Innovation No. S 01/2018*; Fraunhofer ISI: Arlsruhe, Germany, 2018.
99. Reuter, B.; Hendrich, A.; Hengstler, J.; Kupferschmid, S.; Schwenk, M. Rohstoffe für innovative Fahrzeugtechnologien. Herausforderungen und Lösungsansätze. 2019. Available online: https://www.e-mobilbw.de/fileadmin/media/e-mobilbw/Publikationen/Studien/Material-Studie_e-mobilBW.pdf (accessed on 16 March 2020).

 *energies*

Article

Demand-Response Application in Wastewater Treatment Plants Using Compressed Air Storage System: A Modelling Approach

Mattia Cottes *, Matia Mainardis, Daniele Goi and Patrizia Simeoni

Department Polytechnic of Engineering and Architecture (DPIA), University of Udine, Via delle Scienze 208, 33100 Udine, Italy; matia.mainardis@uniud.it (M.M.); daniele.goi@uniud.it (D.G.); patrizia.simeoni@uniud.it (P.S.)
* Correspondence: mattia.cottes@uniud.it

Received: 22 July 2020; Accepted: 11 September 2020; Published: 14 September 2020

Abstract: Wastewater treatment plants (WWTPs) are known to be one of the most energy-intensive industrial sectors. In this work, demand response was applied to the biological phase of wastewater treatment to reduce plant electricity cost, considering that the daily peak in flowrate typically coincides with the maximum electricity price. Compressed air storage system, composed of a compressor and an air storage tank, was proposed to allow energy cost reduction. A multi-objective modelling approach was applied by analyzing different scenarios (with and without anaerobic digestion, AD), considering both plant characteristics (in terms of treated flowrate and influent chemical oxygen demand, COD, concentration) and storage system properties (volume, air pressure), together with the current Italian market economic conditions. The results highlight that air tank volume has a strong positive influence on the obtainable economic savings, with a less significant impact held by air pressure, COD concentration and flowrate. In addition, biogas exploitation from AD led to an improvement in economic indices. The developed model is highly flexible and can be applied to different WWTPs and market conditions.

Keywords: wastewater treatment; anaerobic digestion; water-energy nexus; demand response; energy consumption optimization; multi-objective model

---

1. Introduction

The energy-water nexus refers to all the processes representing linkages between the water system and the energy sector, including the trade-off of both resources [1]. The energy-water nexus has been given increasing interest worldwide in recent years, due to climate change, augmented global energy demand and significant water scarcity [2]. With the increasing renewable energy generation, the electricity supply becomes more and more variable, necessitating flexible energy demand sources, able to adapt to supply variability providing demand response (DR) [1]. DR has been defined as "changes in electric use by demand-side resources from their normal consumption patterns in response to changes in the price of electricity, or to incentive payments designed to induce lower electricity use at times of high wholesale market prices or when system reliability is jeopardized" [3].

Wastewater treatment is an energy-intensive sector and its energy demand is continuously augmenting due to the stringent requirements on treated water effluent quality, requiring advanced technologies for pollutant removal [4]. The energy consumption in wastewater treatment accounts for about 3.4% of the total electricity consumption in the United States, being the third largest electricity consumer [5]. Nowadays, the energy utilization and optimization in the wastewater treatment plant (WWTP) sector has become a growing concern, considering both the economic and environmental aspects [6]. WWTPs were generally not designed with energy efficiency as a main target. Furthermore, information about energy consumption and energy recovery in WWTPs is still unsatisfactory in the scientific literature [6].

Wastewater characterized by higher internal energy (in terms of influent chemical oxygen demand (COD) concentration and flow rate) leads to a more consistent energy consumption in the plant, together with an increased sludge production and an augmented bioreactor footprint [6]. Proven techniques that help to improve plant energy balance include anaerobic digestion (AD) [7,8], sludge incineration, photovoltaic generation and thermal energy recovery [6]. Energy recovery from wastewater treatment can help to reduce the overall economic costs and the related environmental impact [4], following sustainability and circular economy principles. Indeed, wastewater can contain up to 12 times the energy that is needed for its treatment. Following a paradigm change, wastewater can be seen no longer as a waste but as a resource, from which nutrients and valuable compounds can be recovered [9]. In addition, the energy surplus from wastewater treatment can be integrated into energy distribution systems, supplying external consumers [10].

DR was often modelled as a portion of the total system demand, that can be shifted from peak hours to off-peak hours [1]. In industrial process models, typically the energy prices are taken as an input and the DR potential is evaluated by comparing different electricity tariff schemes, adopting a model-based or data-driven approach [1]. The model-based approach involves the detailed description of the system performances, with thermodynamic and kinetic aspects [11]. However, considering that the resulting models often include a set of non-linear differential equations, the optimization models typically focus on process electricity demand, abstracted from the physical details. Considering WWTPs, wastewater flow-rate shows a daily pattern that coincides with the electricity demand pattern, having one peak in the late morning and another one in the early evening. Consequently, the electricity demand is high when the system demand is high. A treatment shift from peak to off-peak periods (for example from evening to night-time) could yield significant electricity expenditure savings [1].

The biological secondary treatment in WWTPs typically involves activated sludge (AS) technology, which is the most widely applied treatment worldwide to remove biodegradable carbon, suspended solids (SS) and nutrients from wastewater [12]. The available process models of the biological phase include the activated sludge model number 1 (ASM1), developed by International Water Association (IWA), which describes the biochemical processes within the aerated tank with 8 processes and 13 state variables [13]. Recent studies demonstrated that modelling diurnal energy prices variation by coupling the ASM1 model with an energy pricing and a power consumption model could enable the WWTP managing utility to reduce plant energy consumption [14]. However, as noted by some authors, a joint representation of the electricity system behavior and the wastewater treatment process in an integrated energy system has not been implemented at present. Nowadays, mathematical modelling and simulation tools are being increasingly applied to WWTP upgrading and optimization [15]: multi-objective optimization models, in particular, allow one to account for different objective functions, with the aim of optimizing design and operations of a selected process [16].

Wastewater treatment consumes 0.5–2 kWh electric energy per m^3 of treated water, depending on the selected technologies and plant scheme [17]. The AS biological process is the highest electricity consumer in WWTPs (10.2–71% of total plant electricity consumption) [1], due to the need for continuously supplying oxygen to the basins, sustaining the aerobic degradation of the organic matter. In medium- and large-scale plants, AS systems account for 50–60% of the total electricity need, followed by the sludge treatment (15–25%) and recirculation pumping (15%) sections [18]. Research has actually focused on aeration optimization: the idea of increasing aeration efficiency by water looping through a piping, including a venturi aspirator, was recently proposed [19], achieving an aeration efficiency in the range of submerse aerators.

Interactive multi-objective optimization can support the designers when new WWTPs need to be built, but can also improve the performances of existing WWTPs, considering several conflicting criteria and parameters [20]. The multi-objective optimization was recently applied in a number of cases to the water and wastewater sector, with significant outcomes. A goal programming was proposed in [21] to optimize industrial water networks, by using a mixed-integer linear programming as a very reliable a priori method, considering several antagonist objective functions, such as freshwater flow-rate, number of connections, total energy consumption. A fuzzy goal programming was instead investigated in [22], in order to optimize wastewater treatment by considering different energy costs, pollutant load, influent and effluent concentrations: the proposed model was subsequently applied to a full-scale plant in Spain. A process simulator, modelling wastewater treatment, and an interactive multi-objective optimization software, were studied in [20] as a practically useful tool in plant design and improvement; successively, the simulator was applied to a municipal WWTP. The control strategy optimization proved to be effective also to reduce greenhouse gas (GHG) emissions from wastewater treatment in a cost-effective manner, considering also operational costs and effluent concentrations. It was highlighted, in particular, that a meaningful GHG emission reduction can be achieved without relevant plant modifications, even if this can lead to an increase in ammonia and total nitrogen (TN) concentrations in the treated effluent [23].

In this work, DR was applied to wastewater treatment, considering several alternative solutions to improve WWTP energy consumption, with a positive expected outcome on the plant economic balance. A multi-decisional modelling approach was employed to evaluate the technical and economic feasibility of the different analyzed scenarios. Following a preliminary study and considering the lack of existing literature, it was decided to focus on compressed air tank installation to reduce economic expenses for aeration. The tank is filled during low energy demand periods (off-peak), pre-compressing the air for utilization during higher demand (peak load) periods. Recently, compressed air energy systems (CAES) have gained attention, due to their great power range and high energy density, making them an available solution in those contexts where the traditional storage technologies, such as pumped hydroelectric energy storage (PHES), cannot be implemented.

Recent research aimed at improving CAES round-trip efficiency through isothermal processes [24]. In [25], a way to decrease the energy dependency in CAES was investigated, taking advantage of transient flow. Differently from typical utilization of CAES, that includes a power production regulation purpose, as proposed in [26], in the present study the exploitation of a compressed air storage (CAS) system as an oxygen buffer in the wastewater treatment process is investigated. Literature analysis showed a lack of studies on such a dual usage of a CAS system; moreover, the proposed approach helps WWTPs to move toward smart energy systems, increasing the number of outputs from a single source, as suggested by [27].

Considering this general framework, an effort towards a reduction in energy costs in WWTPs is needed. As far as is known by the authors, this is the first study proposing compressed air introduction in WWTPs following DR principles to diminish operating costs for aeration, leading to a significant economic saving in a simple and practical way. The results can be useful for WWTP managing authorities, as well as for researchers. Two main scenarios were investigated in this work. The first scenario will consider the impact of the simple introduction of CAS in the plant, while the second one will forecast the integration of the self-made electricity by the biogas (produced in the AD process) into the compressed air system. The second identified scenario, in accordance with [24], allows higher renewable energy use, contributing to the sustainability perspective. The operating parameters, including plant potentiality (in terms of treated flowrate), influent characteristics (COD concentration), compressed air tank pressure and volume, were considered to evaluate the technical and economic convenience of the proposed solution to reduce WWTP operating costs. The paper is structured as follows. In Section 2, a framework describing the interaction between treatments, costs, and variables is proposed and the related design optimization model is explained. Results are reported in Section 3 for the two investigated scenarios, while discussion follows in Section 4 and conclusions are drawn in Section 5.

2. Materials and Methods

2.1. System Configuration

Based on a load shift approach, a DR and net zero energy (NZE)-oriented system was proposed for WWTPs. As previously introduced, the aeration during biological wastewater treatment accounts for more than 30% of the plant's total energy consumption. Instead of shifting wastewater treatment to periods characterized by lower energy costs (leading to the need for large storage tanks or equalization basins, not always practically feasible) [1], the possibility of storing the air necessary for the biological treatment was considered. This solution (schematically represented in Figure 1) allows one to store energy in a compressed air tank during low energy cost (off-peak) periods, utilizing it when the cost increases (peak periods). Furthermore, the possibility to use the electricity from biogas (locally produced in AD process) by cogeneration was analyzed. This second scenario would allow one to cover a significant share of the process electrical load, while the heat production could be used to entirely cover plant heat demand.

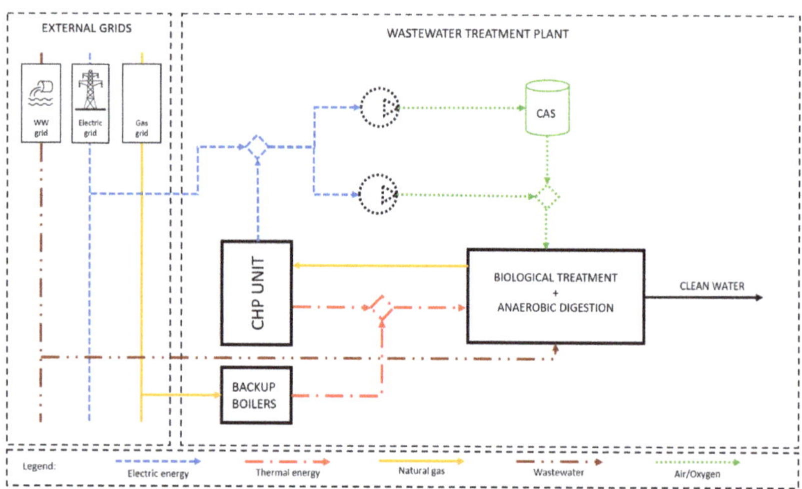

Figure 1. Wastewater treatment plant process flow.

As a first-approach study, the hypothesis of steady state process was made along the whole plant and throughout the different proposed scenarios.

2.2. Input Model Data

The daily pattern of wastewater flowrate was obtained from one-year hourly data of a medium-scale municipal WWTP (86,400 population equivalent (PE), Udine Province), highlighting two distinct peaks at 1 p.m. and 4 p.m. (Figure 2). The obtained data were consistent with the typical reported 24-h behavior of flowrate in municipal WWTPs [1].

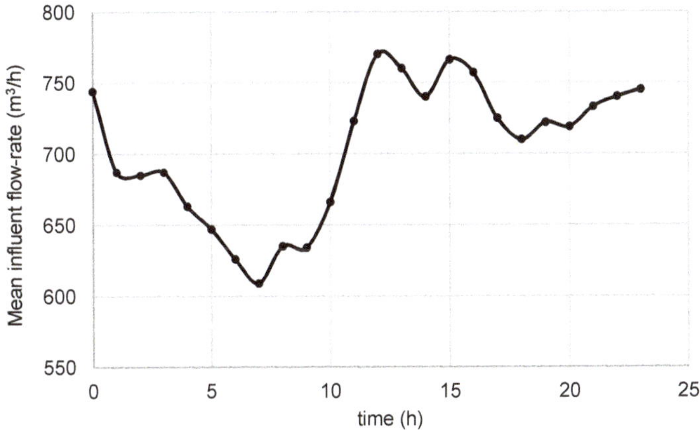

Figure 2. Mean 24-h behavior of the influent flow-rate in the analyzed WW.TP.

The calculated WWTP peak factor (maximum flowrate in the 24 h period divided by mean daily flowrate) was around 1.1, showing a moderate daily variability. The flowrate daily pattern was assumed to be constant for all the analyzed plant potentialities and any seasonal factor was considered in this first approach. The electricity price pattern in the 24 h time period was calculated considering the mean values of a typical month in Italy [28]. The mean electricity price for industrial users was given by the Italian Authority for energy, gas, water and waste [29].

2.3. Biological Treatment Unit

The oxygen injected in the biological phase ($O_{2,cons}$) was supposed to coincide with the oxygen consumed by biomass, estimated by considering a simplified approach (Equation (1)), as proposed in [30]. The operating parameters of the activated sludge process in Equation (1) were set as follows, considering typical full-scale plants in the analyzed territory, where a significant influent dilution is observed in sewers due to aquifer infiltration: mean influent COD concentration (COD_{in}) = 250 mg/L, mean effluent COD concentration (COD_{out}) = 50 mg/L (corresponding to 80% COD abatement), hydraulic retention time (HRT) = 7 h, solid retention time (SRT) = 15 d, biomass concentration in the biological basin (X) = 3.5 g volatile suspended solids (VSS)/L. The factor 1.42 represents COD conversion factor for biomass [30].

$$O_{2,cons} = (COD_{in} - COD_{out}) - \frac{HRT \times X}{SRT} \times 1.42 \tag{1}$$

From the calculated oxygen need and considering oxygen share in the air (20%), it was possible to determine the total injected air flowrate. To evaluate the energy consumption of the biological treatment, the specific operating capacity of oxygen diffusers (fine bubbles) was estimated as 1.2–1.5 kg O_2/kWh (data given from specialized companies in the field). Finally, total plant energy consumption was calculated considering that the electricity consumption for oxygen insufflation is 30–70% of the total plant energy need, as emerged from relevant literature studies [18].

2.4. Compressed Air Storage

The basic idea on which the proposed approach is based is the shift of the peak aeration load to low energy cost periods, introducing a storage system able to sustain the aeration during peak energy cost periods. In order to evaluate the economic convenience of the proposed technology, the energy required to compress the air in the tank, E_{compr} (kJ), is calculated as follows (Equation (2)):

$$E_{compr} = P_{storage} \times V \times ln\frac{P_@}{P_{storage}} + \left(P_{storage} - P_@\right) \times V \tag{2}$$

where V represents the storage tank volume (m^3), $P_{storage}$ is the pressure inside the tank (MPa), and $P_@$ (MPa) is atmospheric pressure.

2.5. Anaerobic Digestion Unit

The sludge production ($\dot{m}_{sludge}$) was calculated using Equation (3), considering full-scale WWTPs' characteristics in the investigated area: a specific sludge production of 40 g suspended solids (SS)/m^3 (p_{sludge}) was used for the successive calculations. Biogas yield from excess sludge (Q_{CH4}, (m^3 CH$_4$/h)) was obtained from Equation (4), considering typical specific methane production (Y_{CH4}, 250 NmL CH$_4$/g VS) vs. the concentration of excess sludge (2.3% w/w) in medium-scale local WWTPs [31].

$$\dot{m}_{sludge} = \frac{Q \times p_{sludge}}{\%VS \times 10^6} \tag{3}$$

$$Q_{CH4} = \dot{m}_{sludge} \times \%VS \times Y_{CH4} \tag{4}$$

The mean electric and thermal efficiencies of the downstream combined heat and power (CHP) unit for biogas cogeneration were assumed respectively as 35% and 43%, consistently with relevant literature studies [32,33].

The heat demand of the anaerobic digester was considered as the sum of the heat losses through digester walls (H_{loss}, (W)) and the thermal energy needed for sludge heating (H_{sludge}, (kJ/h)), calculated using Equations (5) and (6), following the approach proposed in [34]:

$$H_{loss} = k_{air}A_{sup}\left(t_{dig} - t_{air}\right) + k_{soil}A_{base}\left(t_{dig} - t_{soil}\right) \tag{5}$$

$$H_{sludge} = \dot{m}_{sludge}cp_{sludge}\left(t_{dig} - t_{s0}\right) \tag{6}$$

The specific heat capacity of sludge (c_{ps}) was estimated as 3.62 kJ/kg °C [6], while air (t_{air}) and soil (t_{soil}) temperature were obtained from regional climate data [35]. Air (k_{air}) and soil (k_{soil}) heat transfer coefficients were taken from [34]. The digester operating temperature, t_{dig}, was set at 35 °C (optimum mesophilic range [36]), while the mean influent sludge temperature, t_{s0}, was supposed to be 15 °C, consistently with regional climate data [35]. The geometrical characteristics of the digester (base area, A_{base}, (m^2), and lateral area, A_{sup}, (m^2)) were calculated considering the real characteristics of the analyzed full-scale reactor.

2.6. Economic Analysis

For the economic analysis, it is necessary to evaluate the capital cost of the introduced technology and the reduction in operating costs that can be achieved. The capital costs for compressed air storage installation were calculated using the data reported in [37]. A linear correlation between the installed compressor power (W) and the compressor cost (C), obtained through specialized surveys, was considered in this basic approach. Equations in the following form were obtained for both components (compressor and storage tank):

$$C_i = W_i \times c_i + q \tag{7}$$

The total investment cost (C_{TOT}, (€)) is defined as the sum of the capital costs of the installed components. The revenues are defined as the avoided cost for energy purchase, that comes from the load shift due to the compressed air storage. The reduction in operating costs is consequently defined as the difference between the energy cost (purchased in the scenario without air storage) and the effective cost with storage implementation:

$$R_{ope} = C_{av,CHP} + C_{av,tank} + R_{TEE} - C_{compr} - C_{O\&M} \tag{8}$$

In Equation (8), $C_{av,CHP}$ (€/y) is the avoided cost through biogas utilization, $C_{av,tank}$ (€/y) is the avoided cost through the air storage tank, C_{compr} (€/y) is the cost to compress the air inside the tank, and $C_{O\&M}$ (€/y) is the operation and maintenance system cost. R_{TEE} (€/y) is the share of revenues coming from primary energy saving, due to biogas exploitation from AD. CHP-related variables are set to 0 in the scenario where AD is not considered (Scenario 1). An average economic value of current Italian White certificates, equal to 250 EUR/ton of oil equivalent (toe), has been considered in this basic economic analysis for biogas valorization.

The main economic parameter used for the analysis is the net present value (NPV, (EUR)) calculated with a discount rate of 5.5%.

Another parameter used is the pay-back time (PB, (y)) of the investment (Equation (9)), defined as the total investment cost divided by the revenues obtained from the reduced energy cost:

$$PB = \frac{C_{TOT}}{R_{ope}} \tag{9}$$

2.7. Optimization Model Implementation

The multi-objective optimization model was written in MATLAB® and successively implemented in mode FRONTIER® for scenario analysis. The selected parameters for the optimization process are summarized in Table 1. The range of modelled COD concentration was in line with the observed COD values in the analyzed territory, characterized by consistent wastewater dilution, due to mixed sewers and significant infiltrations from the aquifer. A wide range of treated flowrates was considered, to extend the validity of the obtained results to different scale WWTPs. The modelled flowrate range in WWTPs approximately corresponds to plant potentialities of 75,600–225,600 PE, consistent with most of medium scales WWTPs in Friuli-Venezia Giulia region (North-east of Italy). As for compressed air system characteristics, different combinations of pressure and volume were considered, to evaluate the influence of these parameters (both singularly and in combination) on the economic output.

Table 1. Selected input parameters considered in the multi-objective optimization model.

Parameter	Selected Range	Modelling Step
Influent flowrate (m³/h)	504 ÷ 1504	100
Influent COD concentration (mg/L)	150 ÷ 500	50
Electricity consumption for aeration (% of total)	30 ÷ 80	10
Compressed air tank volume (m³)	0.3 ÷ 1000	Variable
Compressed air pressure (Mpa)	0.1 ÷ 35	Variable

Separate simulations were conducted for scenario 1 (CAS integration without AD) and scenario 2 (CAS integration with AD). The relevant output parameters that were considered in the analysis were common economic indexes (PB, net present value (NPV)) and, in the case of scenario 2, meaningful environmental aspects (primary energy saving, PES). Since the main goal of this research was the implementation of a new technology in a WWTP following the DR perspective, the multi-objective optimization functions have been selected among those most representative from a sustainable perspective (Table 2). For the assessment of the economic advantages obtained through this intervention, the NPV maximization objective has been chosen. Regarding the environmental impact, the most meaningful objective was PES maximization.

Table 2. Selected output optimization functions considered in the multi-objective optimization model.

Parameter	Objective Function
NPV (EUR)	Maximization
PES (toe)	Maximization

According to the stakeholders' interests, a constraint of 10 years has been applied to the PB output value.

3. Results

As previously introduced, two main scenarios were investigated, the first one involving only introduction of the air storage unit, composed of an air compressor and a storage tank, to allow air accumulation during the off-peak periods and air utilization from the air tank during the peak periods for sustaining aeration in the biological basin. The second approach includes AD introduction, with electricity production in a CHP unit and integration of the produced electricity in the storage system to increase both total energy savings and the use of renewable energy sources. This solution is particularly indicated in medium and large-scale plants, where AD is typically already implemented.

The simulations were carried on as explained before using a 32 GB RAM, i7 4770 3.40 GHz PC. A population of 500 individuals and 250 generations was adopted, resulting in 125,000 total evaluated designs for both scenarios, sufficient to obtain the convergence of the process in about 2–3 h.

3.1. Scenario 1: Compressed Air Storage without Anaerobic Digestion

In Figure 3, the influence of the selected input parameters (Table 1) on the economic convenience of the proposed storage system is summarized. It can notice that air tank volume has a strong impact on NPV, with a significant increase in the economic income as the volume increases (up to 1000 m^3). The air storage pressure is shown to have a limited influence on NPV, with a slight increase in the economic convenience as a higher air pressure is selected. As for wastewater characteristics, the influent COD concentration and the wastewater flowrate have a mild effect: a more polluted effluent (meaning a higher internal energy) and a higher plant potentiality are slightly favorable for storage tank installation. Moreover, a linear behavior was encountered by analyzing NPV variation with respect to the influent COD concentration, for the wide chosen range of wastewater flowrates. In the most favorable conditions, PB time was lower than 1 y for Scenario 1, highlighting a significant convenience of air storage system installation, given the actual market economic conditions and the investigated plant characteristics.

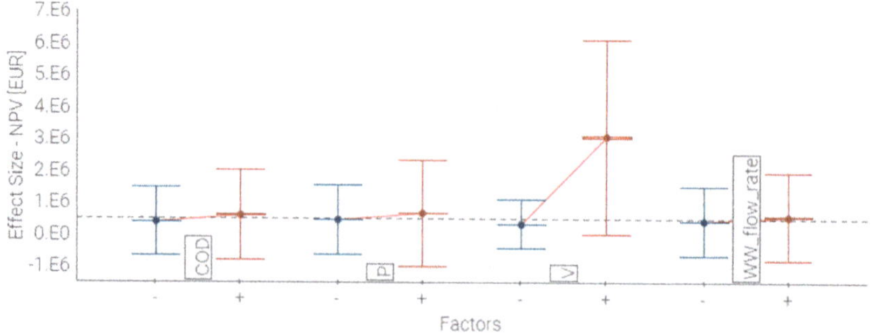

Figure 3. Influence of the main input parameters (wastewater flowrate and influent COD, air storage tank pressure and volume) on the NPV of the proposed compressed air storage system (Scenario 1).

The detailed analysis of NPV variation with respect to air storage pressure, reported in Figure 4, interestingly highlights that NPV increases to a maximum at an intermediate pressure value, while a decrease is observed for a further pressure augmentation (in particular considering larger tank volumes); this is due to the fact that for a higher vessel pressure, the specific compression power increases, leading to higher compression costs.

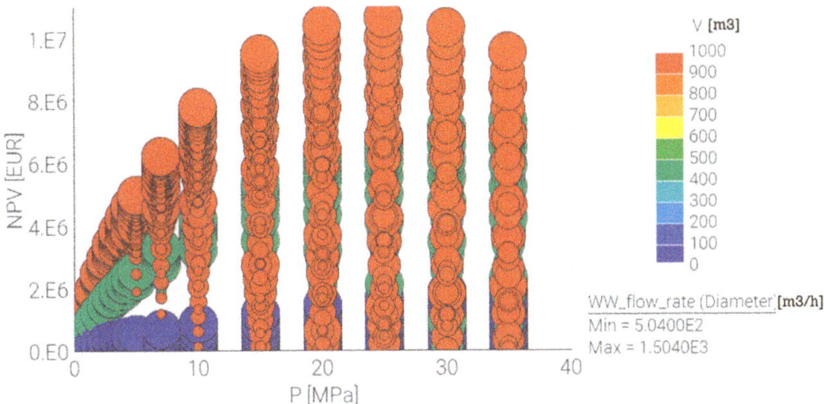

Figure 4. Influence of air storage tank pressure and volume on NPV for different wastewater flowrates (Scenario 1).

3.2. Scenario 2: Compressed Air Storage with Anaerobic Digestion

In the second scenario, AD introduction is included, with biogas valorization through the CHP unit, able to supply a share of the needed electricity. The analysis of the influence of single input parameters on NPV value (Figure 5) shows that a similar behavior to that encountered in Scenario 1 (Figure 3) is obtained, even if influent wastewater flowrate has a stronger influence, due to the fact that AD becomes more convenient for a higher plant potentiality.

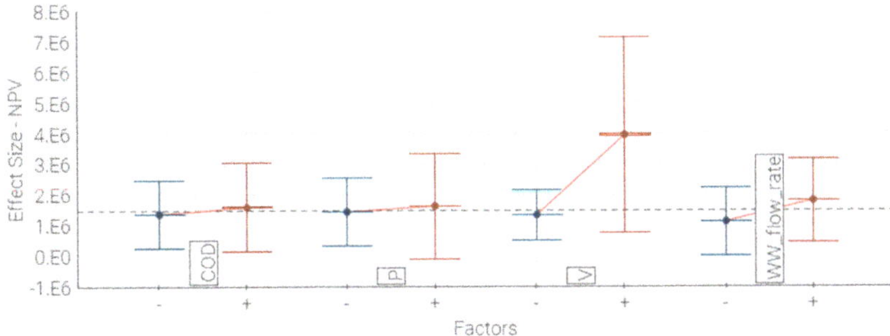

Figure 5. Influence of the main input parameters (wastewater flowrate and influent COD, air storage tank pressure and volume) on the NPV of the proposed compressed air storage system (Scenario 2).

Scenario 2 is particularly representative of medium and large-scale plants, where AD is already applied on full-scale, with biogas valorization: the integration of locally produced electricity with air storage system would optimize energy saving. In Figure 6, NPV and primary energy saving (PES) variation with respect to the influent flowrate was depicted: the maximum obtainable NPV was in the range of EUR 6–10 M, while PES was in the range of 36–108 toe/y for the different analyzed plant potentialities.

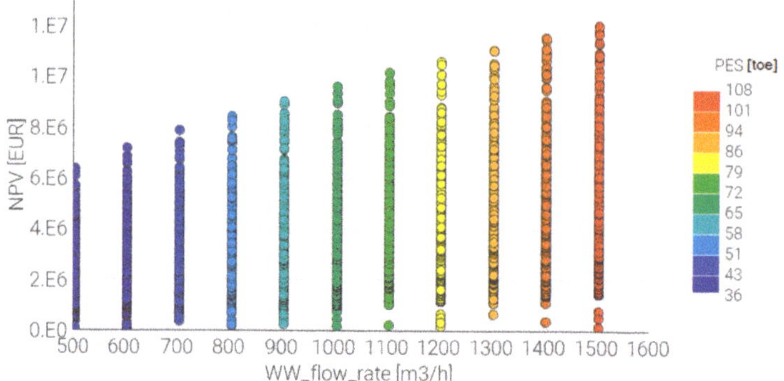

Figure 6. Influence of the treated wastewater flowrate on NPV and PES indices (Scenario 2).

It could be seen that the investigated scenarios had a comparable behavior with respect to the considered input parameters. The proposed CAS system was shown to be technically and economically feasible. Given the cost-effectiveness of the solution, it is now possible to move on to a more in-depth study to analyze the commercially available technical devices for further system optimization.

4. Discussion

Air storage systems have been shown to be particularly useful to allow excess energy storage from renewable energies, with air storage in tanks or caverns (in periods of extra energy production) and a successive expansion of the pressurized air in a turbine (in periods of more consistent energy need). Among the different CAES configurations, including diabatic, adiabatic and isothermal mode, isothermal operations appear to be favorable to increase the overall performances, in terms of system efficiency [24]. The developed model for compressed air storage in WWTPs is highly flexible and can be specifically designed for target WWTPs, specifying plant potentiality and wastewater characteristics, both in terms of influent characterization, energy consumption of the biological phase and operating conditions of the activated sludge basin (such as air diffusers efficiency, HRT, SRT, X). It should be remembered that steady-state conditions, such as the ones considered in the present study, are not typically encountered in daily operations in WWTPs. However, this study was aimed at preliminarily evaluating the feasibility of the proposed solution to reduce operating costs for plant managing authorities. Furthermore, data scarcity has been recently recognized as a limiting factor to allow a widespread utilization of mathematical modelling, considering the complicated usage of online sensors and the huge workload required for sampling campaigns [38].

AD has been largely recognized as a significant source of renewable energy from different organic sources, able to provide fully renewable biogas, that has a high potential energy value [36]. Traditionally, AD was known to be beneficial in large-scale WWTPs, where a sufficient amount of sludge is available to continuously feed the digester, avoiding discontinuous operations throughout the year [39]. However, recently, by analyzing a large range (25,000–1,000,000 PE) of WWTP potentialities, it was proved that AD implementation is convenient also in small-scale plants when assessing environmental and economic benefits, particularly if agro-waste addition can be provided to the digester [40]. This can lead to an enhanced biogas generation, due to an augmented organic load, and a consequently favorable energy balance [40]. Regarding the second proposed scenario, further advances could include the co-digestion of sewage sludge with other locally available organic substrates to increase the obtainable methane yield from AD [31] or the application of sludge pre-treatments to increase its biodegradability [41]. Finally, the possibility to upgrade biogas to high-value biomethane, with technical characteristics totally comparable to fossil-derived natural gas, should be considered [41].

A following phase of the work is forecast, where the available commercial items for air storage will be investigated, to allow for an easier implementation of the proposed optimization system in existing plants. An in-depth process analysis of the aerobic (for example, considering the ASM1 model) and anaerobic (such as Anaerobic Digestion Model Number 1, ADM1 [42]) sections of the plant could be performed, better evaluating the performances of the proposed air storage system in dynamic conditions.

The detailed energy consumption data resulting from WWTP modelling through commercially available software's (such as GPSX®, WEST® or Biowin®) could be integrated in the proposed economic saving approach, with further insights in achievable energy saving.

When considering the oxygen requirement of the biological treatment in WWTPs, it is known that the energy consumption for aeration strongly depends on a pool of different parameters (often difficult to monitor), such as oxygen transfer efficiency, diffuser fouling phenomenon, and diffuser selection [43]. The proposed empirical approach for air supply calculation is extremely simplified and needs to be sustained by an in-depth experimental or modelling campaign. Novel techniques that were recently proposed to save energy in biological wastewater treatment include smart aeration control, consisting of variable frequency drive (VFD), dissolved oxygen (DO) sensors and programmable logic controller (PLC) [44]. When applying the proposed approach to target WWTPs, the experimental data obtained from these novel smart aeration systems could be used to update the input datasets for the proposed economic optimization.

5. Conclusions

The economic optimization of the electricity consumption in existing wastewater treatment plants (WWTPs) following the demand response (DR) principle was proposed in the present study, focusing on the biological treatment, which is known to consume the highest share of electricity in WWTPs. A compressed air storage system, composed of an air compressor and a storage tank, was proposed, considering different operating pressures and volumes. The modelled WWTP characteristics were influent COD concentration and treated flowrate. Two scenarios were analyzed, one with simple air storage integration and the second forecasting the implementation of the air storage system with on-site electricity production from biogas. The results highlight that a short payback time can be obtained considering the actual Italian electricity market conditions, in particular by selecting larger storage volumes. Air storage volume was shown to be the most significant factor affecting the output economic indices, if compared to air pressure and WWTP characteristics. Biogas integration from anaerobic digestion (AD) process would allow one to increase the economic profitability of the investment, with favorable applicability in medium and large-scale WWTPs. A following phase of the work is forecast to study the available devices for commercial air storage systems and air expanders to implement a complete CAES, as well as to study in a deeper manner WWTP dynamic process conditions.

Author Contributions: The authors worked collectively on the manuscript preparation. All authors have read and agreed to the published version of the manuscript.

Funding: This research received no external funding.

Acknowledgments: The authors acknowledge CAFC S.p.A. company for input data for the model. The Company's line of business includes the distribution of water for sale for domestic, commercial, and industrial use, together with wastewater collection and treatment.

Conflicts of Interest: The authors declare no conflict of interest.

Nomenclature

AD	Anaerobic Digestion
ADM1	Anaerobic Digestion Model No. 1
AS	Activated Sludge
ASM1	Activated Sludge Model No. 1
CAS	Compressed air storage
CAES	Compressed air energy storage

CH_4	Methane
CHP	Combined Heat and Power
CO_2	Carbon Dioxide
COD	Chemical oxygen demand
DO	Dissolved Oxygen
DR	Demand Response
GHG	Green-House Gases
HRT	Hydraulic retention time
IWA	International Water Association
NZE	Net Zero Energy
PE	Population Equivalent
PLC	Programmable Logic Controller
TEE	White certificates
TN	Total Nitrogen
VFD	Variable Frequency Drive
VS	Volatile Solids
VSS	Volatile Suspended Solids
WWTP	Waste-Water Treatment Plant

Symbols

% VS	Volatile solid content of sludge (% w/w)
A	Area (m^2)
c	Specific cost (EUR)
C	Cost (EUR])
E	Energy
H	Thermal energy
k	Heat exchange constant
q	Vertical intercept of the generic equation
Q	Flowrate (m^3/h)
R	Revenues (EUR)
t	temperature (°C)
V	Storage tank volume (m^3)
X	Biomass concentration in the biological tank (g VSS/L)
Y_{CH4}	Specific methane yield (NmL CH_4/g VS)

Subscripts and Superscripts

air	air related
base	base digester
dig	digester
sup	superficial digester
av_CHP	avoided through biogas utilization
av_tank	avoided through air storage tank
compr	air compression in the tank related
i	generic device
O&M	operation and maintenance
loss	lost from the digester
in	influent
ope	operating
out	effluent
s0	inlet sludge
sludge	sludge related
soil	soil related
TOT	total investment

References

1. Kirchem, D.; Lynch, M.Á.; Bertsch, V.; Casey, E. Modelling demand response with process models and energy systems models: Potential applications for wastewater treatment within the energy-water nexus. *Appl. Energy* **2020**, *260*, 114321. [CrossRef]
2. Rodriguez, D.J.; Diego, J.; Delgado, A.; DeLaquil, P.; Sohns, A. *Thirsty Energy*; World Bank: Washington, DC, USA, 2013. [CrossRef]
3. Lee, M.P. Assessment of Demand Response & Advanced Metering, Assessment of Demand Response & Advanced Today's Presentation Will Discuss: Purpose of FERC's Annual Assessment Results. December 2012. Available online: http://www.madrionline.org/wp-content/uploads/2013/09/Lee.pdf (accessed on 16 June 2020).
4. Di Fraia, S.; Massarotti, N.; Vanoli, L. A novel energy assessment of urban wastewater treatment plants. *Energy Convers. Manag.* **2018**, *163*, 304–313. [CrossRef]
5. Panepinto, D.; Fiore, S.; Zappone, M.; Genon, G.; Meucci, L. Evaluation of the energy efficiency of a large wastewater treatment plant in Italy. *Appl. Energy* **2016**, *161*, 404–411. [CrossRef]
6. Yang, X.; Wei, J.; Ye, G.; Zhao, Y.; Li, Z.; Qiu, G.; Li, F.; Wei, C. The correlations among wastewater internal energy, energy consumption and energy recovery/production potentials in wastewater treatment plant: An assessment of the energy balance. *Sci. Total Environ.* **2020**, *714*, 136655. [CrossRef]
7. Mainardis, M.; Goi, D. Pilot-UASB reactor tests for anaerobic valorisation of high-loaded liquid substrates in friulian mountain area. *J. Environ. Chem. Eng.* **2019**, *7*, 103348. [CrossRef]
8. Plazzotta, S.; Cottes, M.; Simeoni, P.; Manzocco, L. Evaluating the environmental and economic impact of fruit and vegetable waste valorisation: The lettuce waste study-case. *J. Clean. Prod.* **2020**, *262*, 121435. [CrossRef]
9. Guerra-Rodríguez, S.; Oulego, P.; Rodríguez, E.; Singh, D.N.; Rodríguez-Chueca, J. Towards the implementation of circular economy in the wastewater sector: Challenges and opportunities. *Water* **2020**, *12*, 1431. [CrossRef]
10. Kollmann, R.; Neugebauer, G.; Kretschmer, F.; Truger, B.; Kindermann, H.; Stoeglehner, G.; Ertl, T.; Narodoslawsky, M. Renewable energy from wastewater—Practical aspects of integrating a wastewater treatment plant into local energy supply concepts. *J. Clean. Prod.* **2017**, *155*, 119–129. [CrossRef]
11. Mitra, S.; Grossmann, I.E.; Pinto, J.M.; Arora, N. Optimal production planning under time-sensitive electricity prices for continuous power-intensive processes. *Comput. Chem. Eng.* **2012**, *38*, 171–184. [CrossRef]
12. Tchobanoglus, G.; Burton, F.; Stensel, H.D. *Wastewater Engineering: Treatment and Reuse*, 4th ed.; Metcalf and Eddy: New York, NY, USA, 2003.
13. Henze, M.; Grady, C.; Gujer, W.; Marais, G.; Matsuo, T. Activated sludge model no. 1, Tech. rep. In *IAWQ Scientific and Technical Report No. 1*; IAWQ: London, UK, 1987.
14. Póvoa, P.; Oehmen, A.; Inocêncio, P.; Matos, J.S.; Frazão, A. Modelling energy costs for different operational strategies of a large water resource recovery facility. *Water Sci. Technol.* **2017**, *75*, 2139–2148. [CrossRef]
15. Hvala, N.; Vrečko, D.; Levstek, M.; Bordon, C. The use of dynamic mathematical models for improving the designs of upgraded wastewater treatment plants. *J. Sustain. Dev. Energy Water Environ. Syst.* **2017**, *5*, 15–31. [CrossRef]
16. Simeoni, P.; Ciotti, G.; Cottes, M.; Meneghetti, A. Integrating industrial waste heat recovery into sustainable smart energy systems. *Energy* **2019**, *175*, 941–951. [CrossRef]
17. Gude, V.G. Energy and water autarky of wastewater treatment and power generation systems. *Renew. Sustain. Energy Rev.* **2015**, *45*, 52–68. [CrossRef]
18. Mamais, D.; Noutsopoulos, C.; Dimopoulou, A.; Stasinakis, A.; Lekkas, T.D. Wastewater treatment process impact on energy savings and greenhouse gas emissions. *Water Sci. Technol.* **2014**, *71*, 303–308. [CrossRef]
19. Mahmud, R.; Erguvan, M.; Macphee, D.W. Performance of Closed Loop Venturi Aspirated Aeration System: Experimental Study and Numerical Analysis with Discrete Bubble Model. *Water* **2020**, *12*, 1637. [CrossRef]
20. Hakanen, J.; Miettinen, K.; Sahlstedt, K. Wastewater treatment: New insight provided by interactive multiobjective optimization. *Decis. Support Syst.* **2011**, *51*, 328–337. [CrossRef]

21. Ramos, M.A.; Boix, M.; Montastruc, L.; Domenech, S. Multiobjective Optimization Using Goal Programming for Industrial Water Network Design. *Ind. Eng. Chem. Res.* **2014**, *53*, 17722–17735. [CrossRef]

22. Díaz-Madroñero, M.; Pérez-Sánchez, M.; Satorre-Aznar, J.R.; Mula, J.; López-Jiménez, P.A. Analysis of a wastewater treatment plant using fuzzy goal programming as a management tool: A case study. *J. Clean. Prod.* **2018**, *180*, 20–33. [CrossRef]

23. Sweetapple, C.; Fu, G.; Butler, D. Multi-objective optimisation of wastewater treatment plant control to reduce greenhouse gas emissions. *Water Res.* **2014**, *55*, 52–62. [CrossRef]

24. Castellani, B.; Presciutti, A.; Filipponi, M.; Nicolini, A.; Rossi, F. Experimental investigation on the effect of phase change materials on compressed air expansion in CAES plants. *Sustainability* **2015**, *7*, 9773–9786. [CrossRef]

25. Besharat, M.; Dadfar, A.; Viseu, M.T.; Brunone, B.; Ramos, H.M. Transient-flow induced compressed air energy storage (TI-CAES) system towards new energy concept. *Water* **2020**, *12*, 601. [CrossRef]

26. Abbaspour, M.; Satkin, M.; Mohammadi-Ivatloo, B.; Lotfi, F.H.; Noorollahi, Y. Optimal operation scheduling of wind power integrated with compressed air energy storage (CAES). *Renew. Energy* **2013**, *51*, 53–59. [CrossRef]

27. Dincer, I.; Acar, C. Smart energy systems for a sustainable future. *Appl. Energy* **2017**, *194*, 225–235. [CrossRef]

28. Gestore dei Mercati Energetici (GME). Electricity Market. 2020. Available online: https://www.mercatoelettrico.org/it/ (accessed on 20 February 2020).

29. Autorità di Regolazione Energia Reti e Ambiente (ARERA). Electricity Prices for Industrial Users. 2019. Available online: https://www.arera.it/it/dati/eepcfr2.htm (accessed on 21 February 2020).

30. Dionisi, D.; Rasheed, A.A. Maximisation of the organic load rate and minimisation of oxygen consumption in aerobic biological wastewater treatment processes by manipulation of the hydraulic and solids residence time. *J. Water Process Eng.* **2018**, *22*, 138–146. [CrossRef]

31. Cabbai, V.; Ballico, M.; Aneggi, E.; Goi, D. BMP tests of source selected OFMSW to evaluate anaerobic codigestion with sewage sludge. *Waste Manag.* **2013**, *33*, 1626–1632. [CrossRef]

32. Simeoni, P.; Nardin, G.; Ciotti, G. Planning and design of sustainable smart multi energy systems. The case of a food industrial district in Italy. *Energy* **2018**, *163*, 443–456. [CrossRef]

33. Mainardis, M.; Flaibani, S.; Mazzolini, F.; Peressotti, A.; Goi, D. Techno-economic analysis of anaerobic digestion implementation in small Italian breweries and evaluation of biochar and granular activated carbon addition effect on methane yield. *J. Environ. Chem. Eng.* **2019**, *7*, 103184. [CrossRef]

34. Zupančič, G.D.; Roš, M. Heat and energy requirements in thermophilic anaerobic sludge digestion. *Renew. Energy* **2003**, *28*, 2255–2267. [CrossRef]

35. Osservatorio Meteorologico Regionale (Osmer). Data Archival. 2019. Available online: https://www.osmer.fvg.it/archivio.php?ln=&p=dati (accessed on 26 February 2020).

36. Mainardis, M.; Buttazzoni, M.; Goi, D. Up-Flow Anaerobic Sludge Blanket (UASB) Technology for Energy Recovery: A Review on State-of-the-Art and Recent Technological Advances. *Bioengineering* **2020**, *7*, 43. [CrossRef]

37. Zakeri, B.; Syri, S. Electrical energy storage systems: A comparative life cycle cost analysis. *Renew. Sustain. Energy Rev.* **2015**, *42*, 569–596. [CrossRef]

38. Borzooei, S.; Amerlinck, Y.; Abolfathi, S.; Panepinto, D.; Nopens, I.; Lorenzi, E.; Meucci, L.; Zanetti, M.C. Data scarcity in modelling and simulation of a large-scale WWTP: Stop sign or a challenge. *J. Water Process Eng.* **2019**, *28*, 10–20. [CrossRef]

39. Misson, G.; Mainardis, M.; Incerti, G.; Goi, D.; Peressotti, A. Preliminary evaluation of potential methane production from anaerobic digestion of beach-cast seagrass wrack: The case study of high-adriatic coast. *J. Clean. Prod.* **2020**, *254*, 120131. [CrossRef]

40. Arias, A.; Feijoo, G.; Moreira, M.T. What is the best scale for implementing anaerobic digestion according to environmental and economic indicators? *J. Water Process Eng.* **2020**, *35*, 101235. [CrossRef]

41. Borzooei, S.; Campo, G.; Cerutti, A.; Meucci, L.; Panepinto, D.; Ravina, M.; Riggio, V.; Ruffino, B.; Scibilia, G.; Zanetti, M. Optimization of the wastewater treatment plant: From energy saving to environmental impact mitigation. *Sci. Total Environ.* **2019**, *691*, 1182–1189. [CrossRef]

42. Parker, W.J. Application of the ADM1 model to advanced anaerobic digestion. *Bioresour. Technol.* **2005**, *96*, 1832–1842. [CrossRef]

43. Drewnowski, J.; Remiszewska-Skwarek, A.; Duda, S.; Łagód, G. Aeration process in bioreactors as the main energy consumer in a wastewater treatment plant. Review of solutions and methods of process optimization. *Processes* **2019**, *7*, 311. [CrossRef]

44. Khatri, N.; Khatri, K.K.; Sharma, A. Enhanced Energy Saving in Wastewater Treatment Plant using Dissolved Oxygen Control and Hydrocyclone. *Environ. Technol. Innov.* **2020**, *18*, 100678. [CrossRef]

Article

The Energy Trade-Offs of Transitioning to a Locally Sourced Water Supply Portfolio in the City of Los Angeles

Angineh Zohrabian and Kelly T. Sanders *

Department of Civil and Environmental Engineering, University of Southern California, Los Angeles, CA 90089, USA; azohrabi@usc.edu
* Correspondence: ktsanders@usc.edu

Received: 1 September 2020; Accepted: 9 October 2020; Published: 26 October 2020

Abstract: Predicting the energy needs of future water systems is important for coordinating long-term energy and water management plans, as both systems are interrelated. We use the case study of the Los Angeles City's Department of Water and Power (LADWP), located in a densely populated, environmentally progressive, and water-poor region, to highlight the trade-offs and tensions that can occur in balancing priorities related to reliable water supply, energy demand for water and greenhouse gas emissions. The city is on its path to achieving higher fractions of local water supplies through the expansion of conservation, water recycling and stormwater capture to replace supply from imported water. We analyze scenarios to simulate a set of future local water supply adoption pathways under average and dry weather conditions, across business as usual and decarbonized grid scenarios. Our results demonstrate that an aggressive local water supply expansion could impact the geospatial distribution of electricity demand for water services, which could place a greater burden on LADWP's electricity system over the next two decades, although the total energy consumed for the utility's water supply might not be significantly changed. A decomposition analysis of the major factors driving electricity demand suggests that in most scenarios, a structural change in LADWP's portfolio of water supply sources affects the electricity demanded for water more than increases in population or water conservation.

Keywords: urban water system; local water supply; water-energy nexus; electricity demand; index decomposition analysis

---

1. Introduction

Transition to a low carbon and sustainable society will involve multi-sectoral and multi-disciplinary approaches to support decision-making. Reducing the emissions associated with energy consumption is an obvious component of any robust greenhouse gas mitigation plan, and the water sector is one of the largest energy loads in most municipal regions, making it a valuable opportunity for greenhouse gas reductions [1]. Energy is needed to source, convey, treat and distribute water to residential, commercial and industrial users. Energy is also consumed for wastewater collection and treatment, in order to ensure the safe discharge of treated wastewater effluent into the environment [2]. Based on a 2013 study, about 69 billion kWh or 2% of total electricity consumption in the U.S. was consumed for drinking water supply and wastewater management systems [3]. In some water-stressed regions where local freshwater is not abundant, water systems can consume much more energy than average through large pumping projects or advanced treatment of degraded water sources. California, for example, depends on large pumping networks to deliver raw water from where it originates to large water demand regions in Southern California. Consequently, in California, roughly 7.7% of total electricity was consumed in the water sector in year 2001 based on a study published in 2010 [4].

The energy requirements of water supplies are expected to increase in regions with expected population growth, stricter environmental regulations, increasing water scarcity, growing groundwater depletion, and higher dependence on long-distance inter-basin transfers [5], particularly in areas facing extreme and prolonged droughts [6,7]. Several water-stressed regions and cities have formed initiatives to address the rising challenges of reliable water supply and climate resiliency. In arid and semi-arid parts of the U.S. (such as California and Arizona) and across the world (such as in Israel, Australia, and Saudi Arabia), water utilities have promoted programs to expand water use efficiency, water conservation, water reuse, and other alternative supply options in efforts to mitigate water stress [8–11]. The literature underscores the importance of evaluating the energy tradeoffs of these strategies, particularly in densely populated urban areas. For instance, a multi-sector systems analysis by Bartos and Chester [12] found that water conservation policies in Arizona could reduce statewide electricity demand up to 3%. In another case study for Mumbai [13], a scenario-based approach was used to evaluate the residential water-energy nexus for achieving the Sustainable Development Goals over the time frame of 2011–2050. This study found that the interactions between water and energy during end use (i.e., a change in energy consumption prompts a change in water consumption and visa versa) significantly affected water demand and therefore, the electricity consumption for water supply and wastewater systems [13]. Another study used a cost-abatement curve method to analyze energy and water efficiency opportunities across household appliances, and found that an average U.S. household could annually save 7600 kWh of energy (electricity and natural gas) and 39,600 gallons of water if baseline appliances were replaced by energy and water-efficient appliances [14].

Quantifying the energy and emissions footprint of alternative water supply options has also been studied. A set of key performance indicators were used to compare the performance of six decentralised and three centralised water reuse configurations for the cities of San Francisco del Rincon and Purisima del Rincon in Mexico [15]. The results indicated that decentralised water reuse strategies performed the best in terms of water conservation, greenhouse gas emissions, and eutrophication indicators; however, almost negligible energy savings were reported [15]. Two studies estimated the future energy requirements of urban water management for the City of Los Angeles [16] and Los Angeles county [17]. Both studies highlighted that conservation and alternative supply options could reduce the overall energy consumed for water while an increased reliance on long-distance transfers could exacerbate future energy needs. Another study used a spatially explicit life-cycle assessment method to estimate the emissions associated with different water sources for Los Angeles and concluded that the greenhouse gas emissions footprint of water recycling could be as high as water supplied from some imported sources [18].

This paper builds on the prior literature analyzing the energy trade-offs of the City of Los Angeles's water supply portfolio by estimating the energy and emissions trade-offs of LA's future water supply trajectories through 2050 for a variety scenarios, including those that significantly increase locally sourced water supplies. We analyze how electricity demand and emissions are shifted in time and space across electricity serving utilities in California. We first provide an overview of the city's baseline water supply, as well as a series of projected business-as-usual and local water supply trajectories; second, we estimate the electricity demanded for water across the time frame extending from 2020 through 2050 and identify the main driving factors that affect water-related electricity demand for each trajectory; third, we spatially disaggregate each electricity demand estimate according to the utility delivering electricity for sourcing and/or treating each water source; and finally, we discuss the energy and emissions burden of future water supply trajectories. Our analytical framework is applied to reveal the potential tensions that could arise in efforts to simultaneously increase Southern California's local water supply, while ramping up efforts to decrease greenhouse gas mitigation strategies.

2. Water Supply System of the Los Angeles Department of Water and Power

The water supply system in Los Angeles was engineered in the early twentieth century [19,20]. The Los Angeles Department of Water and Power (LADWP) manages the City's water supply and is the second largest municipal water utility in the U.S. [21], delivering water to nearly four million people living in its service territory [7]. Approximately 560 million cubic meters of water is consumed annually by over 680,000 residential and business water service connections [21]. Much of LADWP's water supply is imported from sources outside the City, as local water supplies are limited and precipitation averages only about 12–15 inches per year [20]. Therefore, a large fraction of LADWP's water portfolio has historically been purchased from Metropolitan Water District (MWD), which pumps water hundreds of miles from the Colorado River (via Colorado River Aqueduct or CRA) and northern parts of the state through California Aqueduct in the State Water Project (SWP-East branch and SWP-West branch). In addition, the City of Los Angeles owns the gravity-fed Los Angeles Aqueduct (LAA), which conveys water from the Owens River in the Eastern Sierra Nevada Mountains to Los Angeles. These aqueducts are shown in Figure 1. During 2012–2016, these three major sources (LAA, SWP, and CRA) collectively served about 84% of LADWP's consumed water [21]. Local groundwater makes up most of the remaining supply. Recycled water in the past few years has offset some non-potable water demand (i.e., for industrial and irrigation uses). Efficiency and conservation have also been major priorities for LADWP because of the limits of its local water supply commensurate with its population. In fact, based on a comparative study [22], LA's success with conservation measures has led to constant reduction in LA's daily per person water use even to levels less than many other major cities in the U.S. and across the world. Additionally during drought periods, LADWP has used mandatory water conservation ordinances to ease water shortages [20,23].

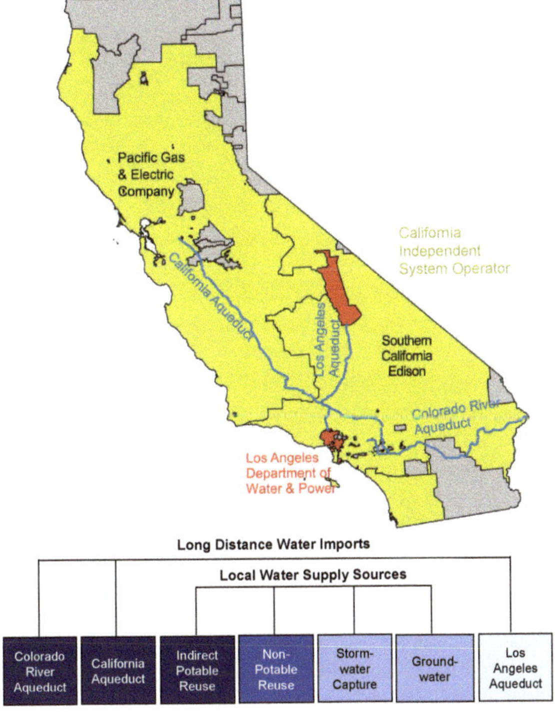

Figure 1. Water supply sources for the Los Angeles Department of Water and Power (LADWP). The areas shown in red and green represent the LADWP and the CAISO (the California Independent System Operator) regions, respectively. The color of each block in the bottom illustration represents the energy intensity of its respective water supply source, where the darkest blue corresponds to the highest energy intensity source and the lightest blue, the lowest.

For Los Angeles, a reliable water supply has been a grand challenge given the region's historical experience with multi-year drought events. Thus, the city seeks alternative sources of water to expand local water availability and to support water supply reliability. Hence, the City of Los Angeles has policy initiatives in its sustainability plan to increase the utilization of local water supplies [24]. These initiatives include:

1. Reducing average per capita potable water use by 22.5% by 2025 and 25% by 2035 compared to the baseline of 133 gallons per capita per day in 2014, as well as maintain or reduce 2035 per capita water use through 2050;
2. Reducing imported water purchases from MWD 50% by 2025 compared to the 2013–2014 fiscal year baseline;
3. Expanding all local sources of water (i.e., groundwater, recycled water, stormwater capture and conservation) to cumulatively account for 70% of the total supply by 2035;
4. Recycling 100% of all wastewater for beneficial reuse by 2035; and
5. Capturing 150,000 acre feet of stormwater, annually, by 2035.

Shifting LADWPs water portfolio will also shift the energy required for its water supply. New water recycling projects can be as energy intensive as MWD imports. (See Figure 1 for the relative energy intensities of LA's local and imported water sources.) Water recycling projects, including non-potable reuse (NPR) and indirect potable reuse (IPR) (via groundwater recharge), are important elements of Los Angeles's plan to increase local water supplies, but they have different energy needs and potential/capacity limitations. Non-potable reuse primarily offsets industrial and irrigation demands [25] (e.g., for agriculture, landscapes, parks, schools, golf courses) and, therefore has limited potential in replacing potable water demands. Furthermore, some industrial facilities have applications that require water that is of higher quality than non-potable water quality (i.e., typically tertiary-level treated) and/or might not have access to recycled water distribution networks. LADWP currently has four recycled water service areas with separate distribution networks that collectively delivered about 45 million cubic meters of NPR in fiscal year 2014/2015, from which approximately 84% was consumed for environmental uses (e.g., for dust control, seawater barriers, and other environmental uses), 14% for irrigation, and 1.6% for industrial applications [7].

IPR via groundwater recharge has higher potential in terms of offsetting urban potable water demands [26]. Requirements for indirect potable categories of recycled water use are different from NPR. IPR requires advanced treatment techniques such as microfiltration, reverse osmosis, ozone, biological activated carbon, and/or advanced oxidation that are often more energy intensive than tertiary treatment and disinfection required for NPR applications [27,28]. In addition, groundwater recharge projects need energy for pumping recycled water from its water treatment location to a groundwater spreading basin (i.e., for injecting water into groundwater aquifers), as well as for pumping water back up from an aquifer and transferring it to potable water distribution network. Since new recycling projects within LADWP's service network are still in their planning stages, there is great uncertainty about their energy footprint and water recovery rates. These factors will depend highly on regional topography, existing land use, the distance between recycled water production and spreading basins [29,30], as well as the type of treatment technology and scale of treatment capacity [31,32]. IPR has a large potential for expansion. The largest wastewater treatment plant in Los Angeles (i.e., Hyperion plant) treats about 363 million cubic meters of wastewater annually. Hyperion currently discharges nearly 83% of its treated wastewater effluent to the Pacific Ocean, which could otherwise be treated to a higher quality to produce recycled water [7]. In addition, there are three smaller wastewater treatment facilities in the city and a few others in neighboring cities that could either produce some amount of recycled water now or be retrofitted to do so. In regards to spreading ground capacity to store recycled water, one study estimated that there are about 30 existing spreading basins in the metropolitan Los Angeles region that are generally underutilized outside the winter months (i.e., approximately 12% of their theoretical infiltration capacity is used) [30].

Stormwater runoff from urban areas is another underutilized local water resource that can be used for groundwater recharge or direct use for landscape irrigation. Stormwater generally requires less intensive water treatment than water treated to IPR standards, and hence, requires less energy. Several centralized and distributed rainwater harvesting projects being pursued by LADWP are estimated to have a total volumetric potential between 163 and 178 million cubic meters by 2035 based on conservative and aggressive scenarios, respectively [7].

The trade-offs between water availability, water supply potential, and the energy requirements of different water supply sources challenge the sustainability of long-term water supply plans; thus, these trade-offs must be accounted in the decision-making process. LADWP's Urban Water Management Plan (UWMP) is a comprehensive water management planning document (mandated by the California Department of Water Resources for every urban water supplier that annually delivers over 3.7 million cubic meters (or 3000 acre-feet) of water annually, or serves more than 3000 urban connections [33]) and is updated in every five years. Although the electricity use in California's water sector is substantial, it is generally voluntary for the water agencies to report water-related energy consumption. LADWP reports information about water supply-related electricity use for historical years in its 2015 UWMP [7], and briefly describes electricity demand trajectories for its future water supply plans. However, there are no energy projections to estimate the consequences of the City of Los Angeles' latest water sustainability goals, which are not yet reflected in LADWP's UWMP. This study addresses this knowledge gap by analyzing the factors that are most likely to drive shifts in the electricity needed for future water supply options.

3. Methods

Here we develop an integrated water-energy systems framework that utilizes a top-down approach to estimate electric load projections for LADWP's water network for the reference year (average between 2010–2015) through 2050 in 5-year increments. We also propose a method to study the relative significance of key factors impacting electricity demand for the utility's evolving water supply over time. Methodological details are described in this section.

3.1. Integrated Water-Energy System Framework

A block diagram of LADWP's water supply stages is plotted in Figure 2 to illustrate the water supply system sources (inputs) and discharges (outputs) considered in this analysis. Our control volume includes the stages involved with supplying water (i.e., surface water supply and recycled water systems). Thus, wastewater management stages (i.e., wastewater collection, wastewater treatment and discharge) are excluded from study boundaries because these processes are managed by a separate entity (i.e., the Los Angeles Bureau of Sanitation), but the water cycle stages involving recycled water production (i.e., additional treatment and distribution) are included in the study as they contribute to LADWP's water supply. For each year studied, we utilize energy intensity values (EI_i in kWh/m^3 for each stage of i) and annual water supply volumes (V_j in m^3 per year) from each water source of j to calculate the total annual electricity demand for the system (E^t in kWh per year), using Equation (1):

$$E^t = \sum_i^n \sum_j^m EI_i V_j^t \tag{1}$$

The energy intensity values of the various water supplies and treatment processes applied within this framework are presented in Table 1. When available, we use EI values reflecting those received by a communication with LADWP or from LADWP's UWMP [7]; otherwise we use EI values from literature [17,34,35]. To address issues related to uncertainty in EI values, we provide electricity demand estimates based on a range of EI values that reflect values in the literature. Otherwise, when no ranges were available, we apply ±20% to nominal EI values. These high and low EI value bounds are noted in parentheses in Table 1.

Table 1. Energy intensities of LADWP water supply sources and electricity serving entities.

Water Supply Source	Water Stage	Electricity Supplier	EI in kWh/m3, Nominal (Low, High) [ref]	Notes
Los Angeles Aqueduct (LAA)	Conveyance	LADWP	0 [7]	LAA aqueduct is entirely gravity fed. Hydropower generation is excluded in this analysis. Water is treated at LA Aqueduct Filtration Plant (LAAFP).
	Treatment	LADWP	0.03 (0.02, 0.04) [17]	
State Water Project – West Branch (SWP – West)	Conveyance	Non-LADWP	2.09 [7] (1.7, 2.5) [1]	This water is purchased from Metropolitan Water District. This water is treated in LAAFP and Jensen Treatment Plant. EI represents the weighted average value.
	Treatment	LADWP	0.03 (0.02, 0.04) [17]	
State Water Project – East Branch (SWP – East)	Conveyance	Non-LADWP	2.6 (2.5, 3.7) [17]	This water is purchased from Metropolitan Water District. The listed energy intensity for treatment is the weighted average of the energy intensities for the Weymouth and Diemer filtration plants.
	Treatment	Non-LADWP	0.03 [7] (0.02, 0.04) [17]	
Colorado River Aqueduct (CRA)	Conveyance	Non- LADWP	1.6 [7] (1.6, 1.9) [17]	This water is purchased from Metropolitan Water District. The listed energy intensity for treatment is the weighted average of the energy intensities for the Weymouth and Diemer filtration plants.
	Treatment	Non-LADWP	0.03 [7] (0.02, 0.04) [17]	
Groundwater pumping	Extraction	LADWP	0.5 [7] (0.2, 0.5) [17]	Groundwater-well pumping energy intensity depends on factors such as water level, and pumping efficiencies. Energy use for treatment is neglected.
Captured stormwater for direct use	Collection	LADWP	0	It is assumed that distributed stormwater capture projects are mainly gravity fed with negligible energy needs. This water is used for on-site outdoor demand.
Non-potable reuse (NPR) (for irrigation and industrial use)	Treatment and distribution	LADWP	0.9 [7] (0.3, 1.1) [34] (for imported NPR, 0.5 [7] (0.4, 0.6) [1])	The assumed EI accounts for additional energy consumed for advanced treatment and for pumping to irrigation and industrial consumers. The NPR distribution network is separate from the potable water system. The EI value for NPR imports accounts only for pumping load.
Indirect potable reuse (IPR) via groundwater recharge from water recycling	Treatment	LADWP	0.6 (0.5, 0.7) [1]	It is assumed that IPR needs an advanced level of treatment. The EI values were chosen based on communication with LADWP for a potential IPR project utilizing the Hyperion wastewater treatment facility's effluent. There is uncertainty associated with these EI values because these projects have not been implemented yet.
	Conveyance and injection	LADWP	0.5 (0.4, 0.6) [1]	
	Extraction	LADWP	0.4 (0.3, 0.5) [1]	
	Conveyance	LADWP	1.5 (1.1, 2.0) [1]	
IPR via groundwater recharge from stormwater capture	Capture and transfer	LADWP	0	It is assumed that stormwater is captured and transferred to groundwater spreading basins by gravity. It is also assumed that EI values for water extraction from ground and conveyance to the water supply system are similar to those communicated by LADWP reflecting water recycling IPR projects.
	Extraction	LADWP	0.4 (0.3, 0.5) [1]	
	Conveyance	LADWP	1.5 (1.2, 1.8) [1]	
Water delivery	Distribution	LADWP	0.1 [7] (0.1, 0.2) [1]	All potable water is delivered to end users by a single central water distribution network.
Non-potable reuse (NPR) (for environmental use)	Treatment	LADWP	0.5 (0.3, 0.8) [35]	This water is treated recycled water, which is used for environmental uses. The EI value reflects the energy needs of advanced treatment with nitrification. The lower and higher values correspond to treatment capacities of 100 million and 1 million gallons per day, respectively [35]. It is assumed that this recycled water is transferred by gravity to environmental project locations.
	Transfer	LADWP	0	

[1] The lower and higher EI values are obtained by applying 20% to listed nominal EI values.

Since LADWP's imported water travels long way to arrive to the city, some of LADWP's water infrastructure, including pumping and raw water treatment, is provided electricity by other utilities. Thus, the electricity-supplier for each energy consuming water facility was determined according to its geographic location based on publicly available documents from LADWP (see Table 1). Two distinct tags (i.e., LADWP and non-LADWP) were applied to distinguish the electric loads supplied by LADWP versus those supplied by other neighboring electric utilities, typically within California Independent System Operator (CAISO).

Other assumptions were made to estimate water-related electricity demands. For example, we assumed no losses in water across each individual water supply stage; in other words, the volume of water entering each facility/stage equals the volume of water exiting that stage, which transfers to the subsequent stage that follows. However, water losses (including firefighting and mainline flushing to improve water quality) are accounted for as non-revenue generating water demands in LADWP's projected total water demand. Thus, we do not make further assumptions regarding to potable water lost to the environment. We understand that ignoring water losses may cause an overestimation of electricity demand, but given the low fractions of water losses in LADWP (the real water losses accounted for 3.8% of total supplied water in 2013/2014 [7]) and the fact that most electricity consumed for water supply occurs upstream of the water distribution system, the significance of this potential overestimation of annual electricity consumption is likely small. Additionally, we assume water leaving the treatment stage is potable and is distributed uniformly across LADWP consumers, regardless of the source or location of treatment and consumption. In terms of recycled water, we consider the marginal energy needs of treating the effluent exiting wastewater treatment facilities to meet recycled water standards, as well as the energy needed for recycled water distribution pumping [36]. We also account for the electricity needed for producing recycled water that is used for beneficial reuse (namely for environmental uses), even though this water is eventually discharged into the environment without offsetting end-use water demand.

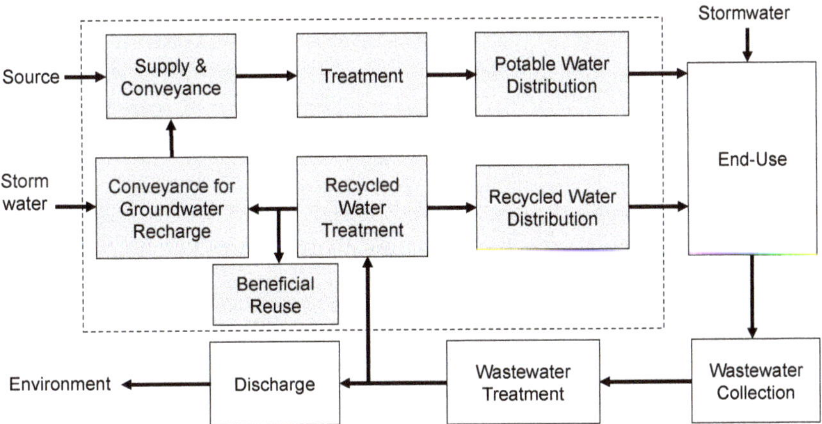

Figure 2. Block diagram of the general components of a water supply system. The dashed box indicates the boundaries of this study.

3.2. Scenario Definitions

Scenarios are developed to explore the energetic tradeoffs of increasing local water supplies over 5-year increments for years spanning 2020 through 2050. We define two scenarios for average weather year conditions (S1, S2), and two additional scenarios (S3, S4) to simulate future water supplies in a single dry year (that assume similar hydrology to 2014/2015), as a proxy for future possible droughts. In these scenarios, S1 and S3 reflect LADWP's most recent water portfolio trajectory from 2020 to 2040 in the utility's 2015 UWMP [7]; however, we extrapolate projections for years of 2045 and 2050. S2 and S4 are developed to simulate a more aggressive local water supply portfolio to represent the newer water policy targets of the Los Angeles City's Green New Deal, which are not reflected in the 2015 UWMP [24]. The scenarios are described in Table 2.

Table 2. Description of scenarios analyzed in this study.

Weather	#	Water Conservation	Water Recycling	Stormwater Capture
Average weather year	S1	Based on assumptions from the LADWP 2015 UWMP [7]	Based on LADWP 2015 UWMP [7]	Based on LADWP 2015 UWMP [7]
	S2	Maximum cost-effective potential based on LADWP water conservation study [37]	Additional recycling from Hyperion wastewater treatment plant	Aggressive potential based on [38]
Single dry year	S3	Based on LADWP 2015 UWMP for single dry year [7]	Based on LADWP 2015 UWMP for single dry year [7]	Based on LADWP 2015 UWMP for single dry year [7]
	S4	Maximum cost-effective potential [37], plus drought-related additional savings	Additional recycling from Hyperion wastewater treatment plant	Conservative potential based on [38]

For S2, we consider the maximum cost-effective conservation potential reported in LADWP's water conservation potential study [37] for the years spanning 2020 and 2035, and we assumed 2% additional conservation for each subsequent five year block thereafter (i.e., for 2040, 2045 and 2050). (Based on LADWP's water conservation potential study [37], cost-effective conservation is defined as the level of water savings achievable through cost-effective conservation programs implemented by LADWP, but it would require customer engagement through expanded financial incentives.) For S4, we assume that water savings in dry years exceed conservation volumes in average weather years due to factors such as more aggressive voluntary and involuntary conservation measures and other water saving ordinances. For stormwater, the cumulative centralized stormwater capture potential reflects LADWP's stormwater capture master plan [38]. To meet the City of Los Angeles' 100% wastewater recycling goal [24], we assume that 60% of the current volume of discharged effluent from wastewater treatment facilities will be further treated according to IPR standards for future groundwater recharge projects by 2035. The remaining 40% is assumed to be treated for environmental use. We exclude any potential water supply from seawater desalination, as LADWP does not include desalinated seawater as part of its future water portfolio [7,16]. Water demand volumes are kept constant in each set of scenarios, such that S1–S2 and S3–S4 reflect water demand volumes in LADWP's 2015 UWMP for an average year and single-dry year, respectively [7]. Figure 3 illustrates LADWP's water portfolio for a historical average year (i.e., ref.), as well as the assumed water portfolios for S1–S4.

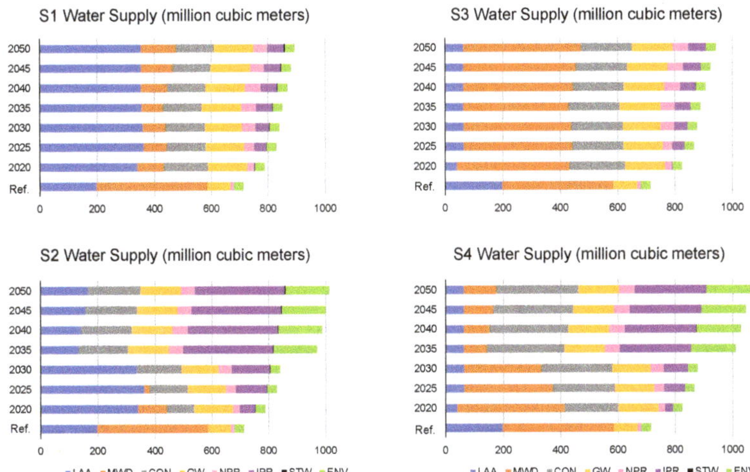

Figure 3. Historical reference year and projected water supply portfolios in S1–S4. Water supplies from LAA (Los Angeles Aqueduct), MWD (Metropolitan Water District), CON (conservation), GW (groundwater), IPR (indirect potable reuse), NPR (non-potable reuse) and STW (stormwater capture for direct use) balance LADWP's historical or projected water demand, and ENV includes recycled water that goes into environmental services.

3.3. Performance Indicators

The percentage of local water supply is defined as the fraction of water demand that is met by cumulative supplies from groundwater, stormwater capture, NPR, IPR, and future conservation. For the projection years (i.e., 2020–2050), volumes of conserved water are included in local water supply percentage calculations because the projected demand for future years does not account for future conservation measures. However, the historical average year is an exception where the water demand value is based on real water consumption data, and therefore, the local water percentage excludes conservation.

Daily water use per person (in the units of liters per person per day, $L/P/D$) is used as an indicator for water demand (note that water demand excludes volume of water for beneficial reuse). Two energy indicators are defined, one to capture the electricity demand intensity of the water system in kWh/m^3, and the other to track annual electricity demand for water supply per person in kWh/P.

We also estimate emissions per unit of water demand in $kgCO_2/m^3$. Total carbon dioxide emissions are estimated using an emissions intensity of 225 $kgCO_2/MWh$, which represents California's 2018 electric grid [39], and 75 $kgCO_2/MWh$, which was applied as a proxy for a future decarbonized grid mix based on [40]. For a future decarbonized grid in 2035, we calculated the average emissions intensity of California's electricity system based on information provided in [40] for the electricity generation fuel mix and total CO_2 emissions in 2030 and 2040. For the state of California to achieve an 80% reduction in California's greenhouse gas emissions by 2050 (from the 1990 levels), the high electrification pathway assumes that the share of renewable energy sources in California's electricity supply is 60% and 79% in 2030 and 2040, respectively [40], which illustrates significant growth from the share of renewables in 2018, i.e., 29% based on [39].

3.4. Decomposing Driving Factors for Electricity Demand

An index decomposition analysis (IDA) is formulated to examine the impact and significance of key factors influencing the electricity demand for LADWP's water system over time. We define influencing factors, including population, water use (influenced by water demand and water conservation programs), and the energy intensity of water supply portfolios. We reformulate Equation (2) to describe the relationship between the three predefined influencing factors and the total electricity demand for the water system:

$$E^t = P^t \times \frac{V^t}{P^t} \times \frac{E^t}{V^t} = P^t \times \frac{V^t}{P^t} \times EI^t = P^t \times \frac{V^t}{P^t} \times V^t \sum_i^n \sum_j^m EI_{i,j} v_j^t \tag{2}$$

where E^t is electricity demand for water, P^t is population, V^t is volume of water supplied (water supplied is equal to projected water demand minus projected conservation), v_j^t is the fraction of water supplied from source j in year t, $EI_{i,j}$ is energy intensity of water supplied from source j in water supply stage i. A change over time in electricity demand can be decomposed into three driving factors relating to the effects of population, physical water supplied per person, and the energy intensity of water supply as it is shown in Equation (3). Here, we apply the most common decomposition method, i.e., additive logarithmic mean Divisia index method I, from [41], to calculate ΔE_P, ΔE_V and ΔE_{EI} shown in Equations (4)–(6), where $L(x,y) = (x-y)/(lnx - lny)$ for $x \neq y$, and $L(x,y) = x$ for $x = y$. We conduct decomposition analysis of total electricity demand change in year 2035 and the reference historical average ($\Delta E = E^{t=2035} - E^{Ref}$), and we repeat the analysis for all four studied scenarios.

$$\Delta E = \Delta E_P + \Delta E_V + \Delta E_{EI} \tag{3}$$

$$\Delta E_P = L(E^t, E^{Ref}) \times ln(P^t / P^{Ref}) \tag{4}$$

$$\Delta E_V = L(E^t, E^{Ref}) \times ln((V^t / P^t)/(V^{Ref} / P^{Ref})) \tag{5}$$

$$\Delta E_{EI} = L(E^t, E^{Ref}) \times ln(EI^t / EI^{Ref}) \tag{6}$$

4. Results

4.1. Electricity Demand

Average annual electricity demand estimates for LADWP's water system from 2020 to 2050 for each scenario are presented in Table 3. For all four scenarios, electricity demand slowly grows over time in almost all projected years between 2020 and 2050. Between 2020–2030, S1 and S2 show a lower electricity demand for water compared to the historical average year. In period between 2035 to 2050, S1 continues to be less energy intensive than the historical average while S2 results in a jump in electricity demand in 2035 due to the large expansion of the IPR supply. The dry weather scenarios (S3 and S4) have higher energy requirements compared to the historical average in almost all studied years, but scenario (S4) has lower electricity demand growth compared to LADWP's 2015 UWMP scenario (S3). In the short-term, electricity demand for S3 and S4 is close to the historical average year while over the long-term, much higher electricity demand is observed. Long-term electricity demand for aggressive local water supply scenarios (i.e., S2, S4) are close in magnitude, despite their differences in hydrology conditions.

More details about the electricity demand and carbon dioxide emissions are presented for the year 2035 in Table 4. The year 2035 is chosen because most water targets are set for that year in the Los Angeles City's Green New Deal plan [24]. Less energy-intensive supplies in S1, such as stormwater capture and aggressive water conservation, reduce water demand and offset energy-intensive MWD imports, such that electricity demand for water in 2035 in this scenario is lower compared to the historical average. By contrast, the S2 aggressive local water supply case in 2035 has total energy requirements that are moderately higher than the historical average (16%). In other words, there are neither significant energy penalties nor energy savings for adopting an aggressive local water supply system. Accordingly, replacing the water pumping loads associated with importing water from MWD, with the energy demands of advanced treatment and pumping, for local water recycling results in a nearly equivalent overall energy footprint. However, the distribution of who provides this electricity for the water supply in each respective scenario changes substantially, and thus has important implications for electric utilities across California. In the high local water supply scenario (S2), water-related electricity provided by LADWP increases over 6 times (from 180 in historical reference year to 1100 GWh in 2035), such that the electricity demand for water grows from approximately 0.8% of total LA's system load in the historical average year to 4% in 2035. At the same time, the electricity that would have otherwise had to be delivered to water pumping infrastructure outside of LADWP's electricity service territory is dramatically reduced as energy intensive imports from Northern California and the Colorado River decrease. Thus, although the amount of electricity consumed in the reference case versus the high local supply case (S2) is similar, the relative fraction of electricity delivered by electric utilities (i.e., LADWP versus other Investor-Owned Utilities in CAISO) shifts dramatically with large electricity demand growth implications for LADWP. While the decarbonized future electric grid significantly reduces the emissions associated with electricity consumption by definition, our analysis indicates that growing water-related electricity demand served by LADWP can increase the total amount of carbon dioxide emissions associated with LADWP's water supply, even with a cleaner future electric grid. But total water-related emissions (i.e., considering the emissions associated with LADWP, as well as other utilities in CAISO) will likely decrease by 2035 compared to historical reference in average weather scenarios because of grid decarbonization.

Table 3. Average total annual electricity demand for LADWP's water supply for the historical reference and projection years between 2020 and 2050 for S1-S4 scenarios. All values are in GWh and are rounded to two significant digits.

Scenario	Ref. Year	Year 2020	Year 2025	Year 2030	Year 2035	Year 2040	Year 2045	Year 2050
S1	780–1200	280–470	370–600	380–620	390–630	420–680	450–730	470–760
S2	780–1200	410–600	390–570	400–580	900–1300	900–1300	900–1300	900–1300
S3	780–1200	790–1200	870–1300	880–1300	880–1400	900–1400	950–1400	980–1500
S4	780–1200	810–1300	800–1400	760–1400	830–2000	850–2000	870–2100	890–2100

Table 4. Annual electrical electricity demand and CO_2 emissions for the reference year and 2035 projections for all four studies scenarios. Note that percentages of water-related electricity demand, calculated in reference to total electricity demands in LADWP and CAISO regions, are calculated based on LADWP's 2017 Retail Electric Sales and Demand Forecast [42] and CAISO demand projections reported in California Energy Demand 2018–2030 Revised Forecast [43], respectively. Electricity demand and emissions values are rounded to two significant digits.

Scenario	Total		LADWP			Outside LADWP		
	Average Electricity Demand (GWh)	CO_2 Emissions (1000 Tonnes)	Average Electricity Demand (GWh)	% LADWP Electric Load	CO_2 Emissions (1000 Tonnes)	Average Electricity Demand (GWh)	% CAISO Electric Load [1]	CO_2 Emissions (1000 Tonnes)
Ref.	970	220	180	0.8%	40	790	0.4%	180
S1 (2035)	530	40–120	380	1%	28–85	150	0.1%	12–35
S2 (2035)	1100	80–250	1100	4%	84–250	0	0%	0–0.1
S3 (2035)	1100	85–250	380	1%	28–85	750	0.3%	56–170
S4 (2035)	1100	80–240	910	3%	68–200	160	0.1%	12–37

[1] We applied an average annual growth rate of 0.84% to CAISO total electricity demand forecast for year 2030 in [43] to estimate total electricity demand for 2035.

The dry year scenarios, S3 and S4, have higher energy needs than the average weather year scenarios due to the limited availability of water from LAA that requires no energy for pumping (which is accommodated by a higher reliance on water supplies from energy-intensive sources). The aggressive water conservation programs implemented in S4 reduced overall water supply needs compared to S3 and hence had lower electricity demands. Considerable amounts of energy were consumed in both S3 and S4 to import water from outside LADWP (i.e., from pumping projects served by Investor Owned Utilities in the CAISO region), but significant electricity demands also occur within LADWP due to increased supply from recycling. Therefore, for S3 and S4 in contrast with S1 and S2, we see a more distributed burden of energy among CAISO and LADWP regions. Higher water-related carbon dioxide emissions are expected for S3 and S4 than S1 but approximately similar to S2. The magnitude of total emissions in S3 and S4 can be lower than historical average if the electric grid decarbonizes significantly by 2035.

4.2. Main Drivers for Electrical Energy Demand

The relationship and relative impact of population growth, water conservation, and water supply mix on electricity demand in the year 2035 is compared to the historical average in Figure 4. This IDA analysis suggests that shifting the water supply portfolio from a more energy intensive system, on average, to a lower energy intensive system is the main driving factor for reducing energy consumption in S1 compared to the reference year. By contrast, transitioning to a water supply mix with more local water supplies in S2 leads to only a slight increase in electricity demand, which means that the average energy intensity of the water supply system in 2035 is slightly increased from the reference historical average. In S1 and S3, the impact of water conservation on reducing electricity demand for water is almost similar to the impact of population growth. In S2 and S4, however, aggressive conservation exceeds the impact of population growth on electricity demand for water. In S3 and S4, electricity demand is slightly higher than the historical average, due to combined effects of aggressive conservation and a shift to more energy-intensive water sources.

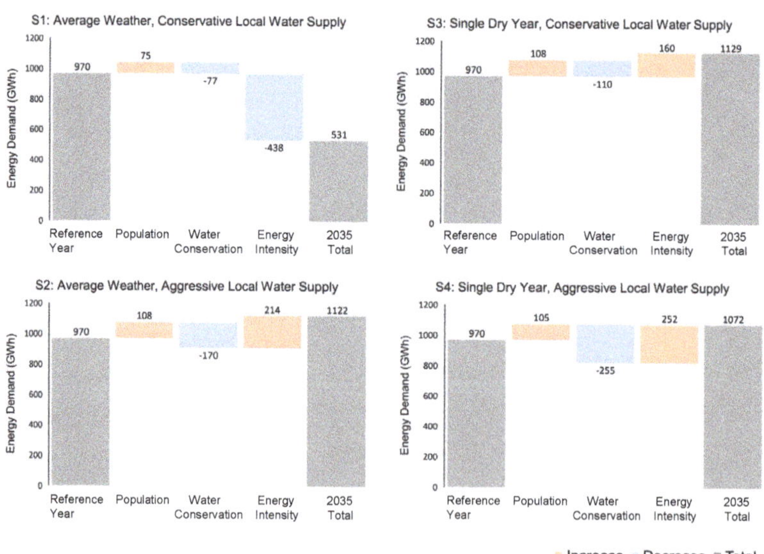

Figure 4. Driving factors for electricity demand changes in 2035 versus the historical average reference year for the four studied water supply scenarios for LADWP.

4.3. Scenario Performance Indicators

Three performance indicators of total annual water-related electricity demand per person in LADWP, annual water use per person per day in LADWP and total electricity demand intensity of water supply (including electricity provided by LADWP and other electric utilities) are illustrated in Figure 5 from the historical reference year to 2020–2050 trajectories. The results show that annual water-related electricity demand per person across all years for LADWP's future water supply scenarios ranges between 98 and 269 kWh/person, with lower values for S1. In fact, the long-term annual electricity demand per person values for S2, S3 and S4 are close to the historical average value (247 kWh/person per year). In all scenarios, whether assuming a dry year or average weather year, the daily water use per person is expected to be lower in all projection years than the historical per person daily water use (475 L/P/D) (due to water conservation assumptions). The large difference between daily water use per person between S2 and S4 is due to the additional conservation that was projected for a future dry weather. The wedges in Figure 5 show the total electricity demand intensity of water supply system. The electricity demand intensity of the water supply system is relatively higher for S2, S3 and S4 (between 1.5 and 2 kWh/m^3) as compared to S1 (between 0.5 to 1 kWh/m^3). The total electricity demand intensity value estimates are comparable to the range of energy intensities reported by Porse et al. [17] as 1.44–1.51 kWh/m^3 and Sanders [16] as 0.53–2.03 kWh/m^3 for average weather condition in 2034/2035.

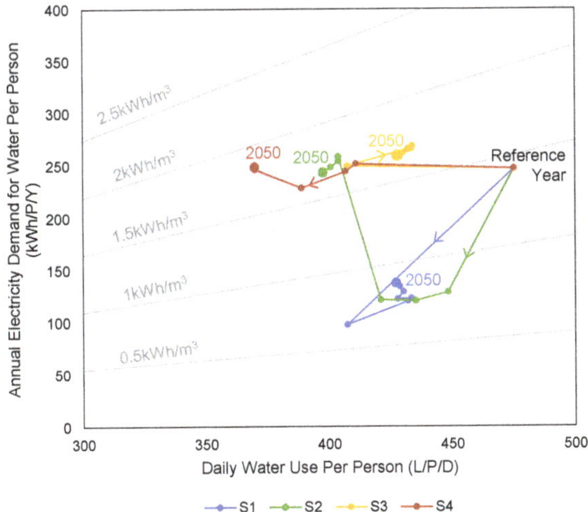

Figure 5. Annual electricity demand per person versus water use per person per day for the studied water supply scenarios for LADWP. The wedges indicate the total electricity demand intensity of water supply system.

5. Discussion

5.1. Energy Trade-Offs of Local Water Supply Options

As water sources dependent on snowpack are vulnerable to drought (e.g., water imports from the Sierra Nevada and the Colorado River), there are large benefits to expanding local water supplies, such as stormwater and recycled water, for mitigating against water shortages during drought. However, increasing recycled water supplies that are treated to potable water quality standards have electricity demand ramifications. Treating water to the quality acceptable for IPR (for groundwater recharge) is generally more energy intensive than treating surface water supplied from distant sources [16], and therefore, there might not be considerable benefits in terms of overall electricity demand for water, and in some cases there may actually be overall energy increases as water recycling projects may need extensive pumping in addition to treatment [6]. In our analysis, we assumed large groundwater recharge projects will be implemented by 2035 in the S2 scenario, when the local water supply percentage will increase substantially at the cost of increased electricity demand in 2035 compared to the previous projection year 2030.

Water demand management strategies, namely water conservation, also have energy impacts. Saving water saves energy that would otherwise be consumed to supply the amount of water saved.

The energy savings associated with each unit of saved water is not equal given LADWP's diverse water portfolio. To reflect LADWP's goals for managing its water supply portfolio over time, we made an underlying assumption that water supply is prioritized from local sources first (which are limited by factors such as treatment and distribution infrastructure for water recycling), then water imports from LAA (which are limited by environmental regulations and hydrology conditions), followed by imports from MWD. Hence, the energy savings of water conservation depends on the marginal supply source that conservation is otherwise avoiding. In this sense, the energy benefits of water conservation are equal to the energy that would otherwise be used importing from MWD, when these water supplies are the marginal source. After the need for MWD purchases is eliminated, the LAA supply becomes the next marginal imported water source, and therefore, the further reduction in water use has lower energy benefits since water from LAA is the least energy intensive source. In other words, the energy benefits of saving water is higher when imports of water from SWP and CRA are the water volumes being displaced, compared to the case in which LAA water is the only source of water imports (i.e., in S2).

Water conservation strategies are not just important for drought resiliency. In lieu of expected growth in population, a continuous commitment to conservation programs is necessary to maintain or even decrease daily per person water use volumes further. Water conservation strategies are not just important for drought resiliency. In lieu of expected growth in population, a continuous commitment to conservation programs is necessary to maintain or even decrease daily per person water use volumes further. Our decomposition analysis suggests that conservation in the average weather scenarios (i.e., S1 and S2) also tends to mitigate increases in energy for water that are driven by higher water demand projections in the year 2035 (compared to the historical average year).

5.2. Spatial Shift in Electricity Demand for Water

Increasing the usage of local water sources in Southern California (and reducing imports from large pumping projects) will cause a dramatic shift in the locations where the water-related electricity demands occur across the state (see Table 3). The large pumping energy requirements of conveying water from the SWP and CRA mostly occur in regions outside of LADWP's electricity service territory; hence, reducing those imports translates in reductions in energy usage by CAISO investor owned utilities including Pacific Gas and Electric and Southern California Edison. On the other hand, the energy loads incurred from the pumping and treatment of local water sources, including groundwater, stromwater and water recycling, are majorly located within LADWP's electricity service territory. Accordingly, moving away from MWD imports and towards local supply sources will shift the energy footprint of water from outside the city into LADWP region. Our spatial disaggregation of electricity demand reveals that transitioning to a local water supply might vastly decrease the electricity demand for LADWP's water supply in the CAISO region. This shift in electricity demand might increase the percentage share of water-related electricity demand from LADWP's total system load from 0.8% in the reference year to about 4% in 2035. This increase in electric load is equivalent to the annual electricity use of over 22,000 average households in California (based on data from [44]). This additional electricity demand in the city will add to its carbon footprint and other upstream environmental externalities associated with electricity generation under current electric grid conditions (although the externalites associated with decreased CAISO generation would be reduced in other regions). The magnitude of environmental externalities will depend on the success of decarbonizing the energy system by moving away from coal and natural gas fuels towards cleaner sources of energy. If the energy transition happens fast enough, the increased electricity demand for water might be insignificant in terms of its carbon footprint. However, with the current electric grid fuel mix, the city's water-related emissions will increase dramatically when aggressive local water supply plans are implemented.

5.3. Achieving Water-Related Sustainability Targets

For comparison, the performance indicators calculated for all four scenarios are summarized in Table 5. The estimated ranges of electricity demand intensity values in 2035 for S1 are lower than the reference historical average, while the other three scenarios show either a similar or a higher range. The emissions intensity of water depends on the assumed average emissions factor of the future electric grid and the electricity demand intensity of the water supply mix. The combination of these two factors show that the emissions intensity could be much lower for a cleaner grid; however, under the current electric grid fuel mix, the emissions intensity of water could be higher in S2–S4 than the historical value (0.3 $kgCO_2/m^3$).

Comparing the performance of studied scenarios with the water targets of the Los Angeles City plan [24] highlights a few points:

- Reducing water purchases from MWD by 50% by the end of 2025 is achievable provided that weather conditions stay normal (S1 and S2). However, even under the aggressive local water supply scenario (S4) in a dry year, achieving this water supply target remains unreachable if the expansion of local water supply is delayed until 2035.
- Supplying 70% of LADWP's water from local resources by 2035 is more likely to be achieved with a more aggressive local water supply expansion (S2 and S4) compared to LADWP's 2015 plans (S1 and S3) that requires unlocking substantial conservation and water recycling potential.
- The 2035 water use per person per day target will be difficult to realize under LADWP's 2015 UWMP, and therefore, will require water savings in addition to reaching its cost-effective conservation potential. With aggressive conservation, LADWP's per person daily water use can position the city among other large cities such as Toronto, Sydney and Melbourne that historically had daily per person water use values lower than LADWP (in the range of 250–435 L/P/D) [22].

Table 5. Summary of performance of studied scenarios in achieving LA's sustainability targets, as well as electricity demand intensity and emissions intensity of water for year 2035 of each scenario.

LA's Green New Deal Targets					
Indicator	Target [24]	S1	S2	S3	S4
Reduction in MWD imports by 2025	50%	83%	96%	21%	36%
% Local Supply by 2035	70%	47%	84%	50%	83%
Per Capita Water Use by 2035 (L/P/D)	379	428	404	428	370
Electricity Demand and CO_2 Footprint					
Indicator	Ref. Year	S1 (2035)	S2 (2035)	S3 (2035)	S4 (2035)
Electricity Demand Intensity (kWh/m^3)	1.1–1.7	0.6–0.9	1.4–2.0	1.3–2.0	1.4–3.4
CO_2 Intensity ($kgCO_2/m^3$) [1]	0.3	0.06–0.2	0.1–0.4	0.1–0.4	0.1–0.4

[1] For the reference year, the value is calculated only based on emissions intensity of California's 2018 electric grid; therefore, no range is provided. The other ranges use two different emissions intensities that are applied to average electricity demand estimates (i.e., 225 and 75 $kgCO_2/MWh$ representing California's 2018 [39] and a future decarbonized electric grid [40], respectively).

5.4. Policy and Planning Implications

The results of this study demonstrate important implications for urban system planning with broader sustainability objectives. The case study of LADWP shows that water policies intended to increase the sustainability and resilience of its water system can be energy intensive and burden its electricity system, thus challenging other sustainability objectives such as greenhouse gas mitigation. Extensive coordination between the water and energy sectors is needed, so that transitions in the energy and water systems facilitate short-term and long-term sustainability priorities without exacerbating tensions between the two sectors.

We recommend that the energy trajectories of future water supply portfolios be incorporated into the city's long-term planning to facilitate holistic decision-making in regards to electricity demand planning, drought resiliency considerations, and ecosystem protections. Synergistic opportunities in water and energy systems open new areas of innovative solutions such as improving the energy efficiency of operations, coordination of water pumping operations with water storage, resource recovery, on-site energy generation, technological innovations, and water trading schemes that could benefit both water and energy sectors, and could reduce the energy needs of water systems [45–51]. Future water systems should be designed based on holistic systems' paradigms that are multipurpose and integrative to promote reliability, resilience and sustainability for the city's urban water system.

6. Conclusions

In this paper, we assess a set of future water supply scenarios for LADWP, which markedly increase local water sources including water recycling, stormwater capture, and conservation in order to support reliable and resilient water supply. We evaluate these scenarios as possible pathways for achieving the city's sustainability water targets using indicators to track the performance of each scenario in terms of electricity demand per unit of water, emissions per unit of water, daily water use per person, and annual electricity demand per person. We estimate the energy requirements for adopting each new water portfolio through 2050 and find that major shifts in water supply sources (i.e., from imports to local supplies) might not significantly change overall energy requirements, but they will change the distribution of the water-related electric load for LADWP's electricity balancing area and other electric load serving entities. Expanding groundwater recharge projects will add additional electric demands on LADWP's electricity system that could be more than 6 times the historical water supply-related electricity demand of LADWP's territory. By contrast, non-LADWP electric load serving entities will likely experience lower electricity demands due to reduced water imports that are transferred over long distances from the city. The decomposition analysis concludes that aggressive conservation measures are important to offset the growth in water demand and the effects of population increases, but the total electricity demand intensity of the water supply has the highest impact on the energy requirements of the water sector in most scenarios.

Our results emphasize that if the City's local water supply adoption occurs at a faster rate than its decarbonization goals in the power sector, those water policies might cause an interim increase in greenhouse gas emissions across LADWP's service territory (due to the increased electricity demand for water) which will be in conflict with the City's goals for mitigating future greenhouse gas emissions. Moreover, this potential growth in electricity demand within LADWP's network might increase other environmental externalities from power generators serving the utility (e.g., increased air pollution and cooling water needs), while non-LADWP regions might benefit environmentally from reduced electricity generation. Thus, evaluating the electricity demands associated with the expansion of a more drought resilient local water supply is important to meeting the multi-faceted sustainability goals of the city, especially those related to clean energy systems.

Author Contributions: Conceptualization, A.Z. and K.T.S.; methodology, A.Z. and K.T.S.; formal analysis, A.Z.; investigation, A.Z.; data curation, A.Z.; writing—original draft preparation, A.Z.; writing—review and editing, A.Z. and K.T.S.; visualization, A.Z.; supervision, K.T.S.; project administration, K.T.S.; funding acquisition, K.T.S. All authors have read and agreed to the published version of the manuscript.

Funding: This research was funded by Los Angeles Department of Water and Power through Los Angeles 100% Renewable Energy Study, grant number ACT-8-LADWP-03.

Acknowledgments: The authors would like to thank the professionals at the Los Angeles Department of Water and Power that helped inform the assumptions utilized in this study.

Conflicts of Interest: The authors declare no conflict of interest. The funders had no role in the design of the study; in the collection, analyses, or interpretation of data; in the writing of the manuscript, or in the decision to publish the results.

References

1. Sanders, K.T.; Webber, M.E. Evaluating the energy consumed for water use in the United States. *Environ. Res. Lett.* **2012**, *7*, 034034. [CrossRef]
2. Capodaglio, A.; Olsson, G. Energy Issues in Sustainable Urban Wastewater Management: Use, Demand Reduction and Recovery in the Urban Water Cycle. *Sustainability* **2019**, *12*, 266. [CrossRef]
3. EPRI. *Electricity Use and Management in the Municipal Water Supply and Wastewater Industries*; Electric Power Research Institute: Palo Alto, CA, USA, 2013; pp. 1–194.
4. GEI Consultants/Navigant Consulting. *Embedded Energy in Water Studies-Study 1: Statewide and Regional Water-Energy Relationship*; California Public Utilities Commission: San Francisco, CA, USA, 2010.
5. Liu, Y.; Hejazi, M.; Kyle, P.; Kim, S.H.; Davies, E.; Miralles, D.G.; Teuling, A.J.; He, Y.; Niyogi, D. Global and Regional Evaluation of Energy for Water. *Environ. Sci. Technol.* **2016**, *50*, 9736–9745. [CrossRef]
6. Hendrickson, T.P.; Bruguera, M. Impacts of groundwater management on energy resources and greenhouse gas emissions in California. *Water Res.* **2018**, *141*, 196–207. [CrossRef]
7. LADWP. *The 2015 Urban Water Management Plan*; Technical Report; Los Angeles Department of Water and Power: Los Angeles, CA, USA, 2015.
8. Almazroui, M.; Şen, Z.; Mohorji, A.M.; Islam, M.N. Impacts of Climate Change on Water Engineering Structures in Arid Regions: Case Studies in Turkey and Saudi Arabia. *Earth Syst. Environ.* **2019**, *3*, 43–57. [CrossRef]
9. Díaz, P.; Morley, K.M.; Yeh, D.H. Resilient urban water supply: Preparing for the slow-moving consequences of climate change. *Water Pract. Technol.* **2017**, *12*, 123–138. [CrossRef]
10. Zhang, X.; Chen, N.; Sheng, H.; Ip, C.; Yang, L.; Chen, Y.; Sang, Z.; Tadesse, T.; Lim, T.P.Y.; Rajabifard, A.; et al. Urban drought challenge to 2030 sustainable development goals. *Sci. Total. Environ.* **2019**. [CrossRef]
11. Morote, Á.F.; Olcina, J.; Hernández, M. The Use of Non-Conventional Water Resources as a Means of Adaptation to Drought and Climate Change in Semi-Arid Regions: South-Eastern Spain. *Water* **2019**, *11*, 93. [CrossRef]
12. Bartos, M.D.; Chester, M.V. The Conservation Nexus: Valuing Interdependent Water and Energy Savings in Arizona. *Environ. Sci. Technol.* **2014**, *48*, 2139–2149. [CrossRef]
13. De Stercke, S.; Chaturvedi, V.; Buytaert, W.; Mijic, A. Water-energy nexus-based scenario analysis for sustainable development of Mumbai. *Environ. Model. Softw.* **2020**. [CrossRef]
14. Chini, C.M.; Schreiber, K.L.; Barker, Z.A.; Stillwell, A.S. Quantifying energy and water savings in the U.S. residential sector. *Environ. Sci. Technol.* **2016**, *50*, 9003–9012. [CrossRef]
15. Landa-Cansigno, O.; Behzadian, K.; Davila-Cano, D.I.; Campos, L.C. Performance assessment of water reuse strategies using integrated framework of urban water metabolism and water-energy-pollution nexus. *Environ. Sci. Pollut. Res.* **2020**, *27*, 4582–4597. [CrossRef]
16. Sanders, K.T. The energy trade-offs of adapting to a water-scarce future: Case study of Los Angeles. *Int. J. Water Resour. Dev.* **2016**, *32*, 362–378. [CrossRef]
17. Porse, E.; Mika, K.B.; Escriva-Bou, A.; Fournier, E.D.; Sanders, K.T.; Spang, E.; Stokes-Draut, J.; Federico, F.; Gold, M.; Pincetl, S. Energy use for urban water management by utilities and households in Los Angeles. *Environ. Res. Commun.* **2020**, *2*, 015003. [CrossRef]
18. Fang, A.J.; Newell, J.P.; Cousins, J.J. The energy and emissions footprint of water supply for Southern California. *Environ. Res. Lett.* **2015**, *10*, 114002. [CrossRef]
19. Pincetl, S.; Newell, J.P. Why data for a political-industrial ecology of cities? *Geoforum* **2017**, *85*, 381–391. [CrossRef]
20. Pincetl, S.; Porse, E.; Cheng, D. Fragmented Flows: Water Supply in Los Angeles County. *Environ. Manag.* **2016**, *58*, 208–222. [CrossRef]
21. LADWP. *Briefing Book 2017–2018*; Los Angeles Department of Water and Power: Los Angeles, CA, USA, 2017.

22. Lam, K.L.; Kenway, S.J.; Lant, P.A. Energy use for water provision in cities. *J. Clean. Prod.* **2017**, *143*, 699–709. [CrossRef]

23. Gonzales, P.; Ajami, N. Social and Structural Patterns of Drought-Related Water Conservation and Rebound. *Water Resour. Res.* **2017**, *53*, 10619–10634. [CrossRef]

24. Mayor Eric Garcetti. *L.A.'s Green New Deal: Sustainable City pLAn*; Technical Report; Mayor's Office: Los Angeles, CA, USA, 2019.

25. Archer, J.E.; Luffman, I.; Andrew Joyner, T.; Nandi, A. Identifying untapped potential: A geospatial analysis of Florida and California's 2009 recycled water production. *J. Water Reuse Desalin.* **2019**, *9*, 173–192. [CrossRef]

26. Thompson, J.; You, E.; VanMeter, H.B.; Chen, C.S.; Goh, A.; Heywood, B. *Groundwater Replenishment Master Planning Report*; Log Aneles Department of Water and Power, and Department of Public Works: Los Angeles, CA, USA, 2012.

27. Cornejo, P.K.; Santana, M.V.; Hokanson, D.R.; Mihelcic, J.R.; Zhang, Q. Carbon footprint of water reuse and desalination: A review of greenhouse gas emissions and estimation tools. *J. Water Reuse Desalin.* **2014**, *4*, 238–252. [CrossRef]

28. Capodaglio, A.G. Fit-for-purpose urban wastewater reuse: Analysis of issues and available technologies for sustainable multiple barrier approaches. *Crit. Rev. Environ. Sci. Technol.* **2020**, *50*, 1–48. [CrossRef]

29. Fournier, E.D.; Keller, A.A.; Geyer, R.; Frew, J. Investigating the Energy-Water Usage Efficiency of the Reuse of Treated Municipal Wastewater for Artificial Groundwater Recharge. *Environ. Sci. Technol.* **2016**, *50*, 2044–2053. [CrossRef] [PubMed]

30. Bradshaw, J.L.; Luthy, R.G. Modeling and Optimization of Recycled Water Systems to Augment Urban Groundwater Recharge through Underutilized Stormwater Spreading Basins. *Environ. Sci. Technol.* **2017**, *51*, 11809–11819. [CrossRef] [PubMed]

31. Paul, R.; Kenway, S.; Mukheibir, P. How scale and technology influence the energy intensity of water recycling systems—An analytical review. *J. Clean. Prod.* **2019**. [CrossRef]

32. Cornejo, P.K.; Zhang, Q.; Mihelcic, J.R. How Does Scale of Implementation Impact the Environmental Sustainability of Wastewater Treatment Integrated with Resource Recovery? *Environ. Sci. Technol.* **2016**, *50*, 6680–6689. [CrossRef]

33. California Department of Water Resources. *2015 Urban Water Management Plans Guidebook for Urban Water Suppliers*; California Department of Water Resources: Sacramento, CA, USA, 2016.

34. Stokes-Draut, J.R.; Taptich, M.N.; Kavvada, O.; Horvath, A. Evaluating the electricity intensity of evolving water supply mixes: The case of California's water network. *Environ. Res. Lett.* **2017**. [CrossRef]

35. Stillwell, A.; Hoppock, D.; Webber, M. Energy Recovery from Wastewater Treatment Plants in the United States: A Case Study of the Energy-Water Nexus. *Sustainability* **2010**, *2*, 945–962. [CrossRef]

36. Stillwell, A.S.; Twomey, K.M.; Osborne, R.; Greene, D.M.; Pedersen, D.W.; Webber, M.E. An integrated energy, carbon, water, and economic analysis of reclaimed water use in urban settings: A case study of Austin, Texas. *J. Water Reuse Desalin.* **2011**, *1*, 208. [CrossRef]

37. Los Angeles Department of Water and Power Briefing. *Executive Report Water Conservation Potential Study*; Los Angeles Department of Water and Power: Los Angeles, CA, USA, 2017.

38. LADWP. *Stormwater Capture Master Plan*; Los Angeles Department of Water and Power: Los Angeles, CA, USA, 2015.

39. EPA. *eGRID Summary Tables 2018*; U.S. Environmental Protection Agency: Washington, DC, USA, 2020.

40. Ming, Z.; Olson, A.; De Moor, G.; Jiang, H.; Schlag, N. *Long-Run Resource Adequacy under Deep Decarbonization Pathways for California*; Energy and Environmental Economics: San Francisco, CA, USA, 2019.

41. Ang, B.W. LMDI decomposition approach: A guide for implementation. *Energy Policy* **2015**, *86*, 233–238. [CrossRef]

42. LADWP. *2017 Retail Electric Sales and Demand Forecast*; Los Angeles Department of Water and Power: Los Angeles, CA, USA, 2017.

43. Kavalec, C.; Gautam, A.; Jaske, M.; Marshall, L.; Movassagh, N.; Vaid, R. *California Energy Demand 2018–2030 Revised Forecast*; California Energy Commission: Sacramento, CA, USA, 2018.

44. EIA. *Household Energy Use in California*; Energy Information Administration: Washington, DC, USA, 2009.

Energies **2020**, *13*, 5589

45. Luthy, R.G.; Wolfand, J.M.; Bradshaw, J.L. Urban Water Revolution: Sustainable Water Futures for California Cities. *J. Environ. Eng.* **2020**, *146*, 04020065. [CrossRef]
46. Achilli, A.; Cath, T.Y.; Childress, A.E. Power generation with pressure retarded osmosis: An experimental and theoretical investigation. *J. Membr. Sci.* **2009**, *343*, 42–52. [CrossRef]
47. Gude, V.G. Energy and water autarky of wastewater treatment and power generation systems. *Renew. Sustain. Energy Rev.* **2015**. [CrossRef]
48. Strazzabosco, A.; Kenway, S.; Lant, P. Solar PV adoption in wastewater treatment plants: A review of practice in California. *J. Environ. Manag.* **2019**, *248*, 109337. [CrossRef] [PubMed]
49. Molinos-Senante, M.; Guzmán, C. Benchmarking energy efficiency in drinking water treatment plants: Quantification of potential savings. *J. Clean. Prod.* **2018**, *176*, 417–425. [CrossRef]
50. Gonzales, P.; Ajami, N.K. Goal-based water trading expands and diversifies supplies for enhanced resilience. *Nat. Sustain.* **2019**, *2*, 138–147. [CrossRef]
51. Daigger, G.T.; Sharvelle, S.; Arabi, M.; Love, N.G. Progress and promise transitioning to the one water/resource recovery integrated urban water management systems. *J. Environ. Eng.* **2019**, *145*, 04019061. [CrossRef]

Publisher's Note: MDPI stays neutral with regard to jurisdictional claims in published maps and institutional affiliations.

MDPI

St. Alban-Anlage 66

4052 Basel

Switzerland

Tel. +41 61 683 77 34

Fax +41 61 302 89 18

www.mdpi.com

Energies Editorial Office

E-mail: energies@mdpi.com

www.mdpi.com/journal/energies